ECOLOGICAL EFFECTS OF AFFORESTATION

Studies in the history and ecology of afforestation in Western Europe

Detail from a map drawn in 1603 of the Taro and Vara valleys showing the characteristic landscape of ancient grassland associated with Appenine commonland (see Chapter 7). The map was drawn by the painter-cartographer Cristofore de'Grassi. Archivo di Stato di Genova *Raccolta Cartografica* n.g. 1177. Aut n. 22/92 (11/11/92).

ECOLOGICAL EFFECTS OF AFFORESTATION

Studies in the history and ecology of afforestation in Western Europe

Edited by

Charles Watkins

Department of Geography,
University of Nottingham,
University Park,
Nottingham, NG7 2RD

C·A·B International

on behalf of the
European Science Foundation

C·A·B International
Wallingford
Oxon OX10 8DE
UK

Tel: Wallingford (0491) 32111
Telex: 847964 (COMAGG G)
Telecom Gold/Dialcom: 84: CAU001
Fax: (0491) 33508

Cataloguing data is available from the British Library

ISBN 0 85198 818 0

Typeset by Leaper & Gard Ltd, Bristol
Printed and bound in the UK by Redwood Press, Melksham

Contents

Contributors

The Editor

Charles Watkins is Lecturer in Geography at the University of Nottingham and was formerly Assistant Director of the Centre for Rural Studies at the Royal Agricultural College, Cirencester, UK. His main research interests are in rural geography, land management and landscape history. Recent books include *Woodland Management and Conservation* (published by David and Charles, 1990). He is co-author of *Justice Outside the City: Access to Legal Services in Rural Britain* (published by Longman, 1991) and *Church and Religion in Rural England* (published by Clark, 1991), and co-editor of *People in the Countryside: Studies of Social Change in Rural Britain* (published by Paul Chapman, 1991).

List of Contributors

N. Allott *Environmental Sciences Unit, Trinity College, Dublin 2, Ireland.*

M. Brennan *Environmental Sciences Unit, Trinity College, Dublin 2, Ireland.*

W. Buysse *Laboratory of Forestry, State University of Ghent, Geraardsbergse Steenweg 267, B 9090 Gontrode, Belgium.*

G.F. Croce *Istituto di Storia Moderna e Contemporanea, University of Genoa, Via Balbi 6, 16126, Genoa, Italy.*

A. Eacrett *Environmental Sciences Unit, Trinity College, Dublin 2, Ireland.*

S.J. Essex *Department of Geographical Sciences, University of Plymouth, Drake Circus, Plymouth, PL4 8AA, United Kingdom.*

M.A. Guido *Istituto ed Orto Botanico 'Hanbury', University of Genoa, Via Balbi 6, 16126, Genoa, Italy.*

M. Guidi *Istituto di Selvicoltura, Universita' di Firenze, Via San Bonaventura 13, 50145, Florence, Italy.*

R. Haines-Young *Department of Geography, University of Nottingham, Nottingham, NG7 2RD, United Kingdom.*

J.C. Iturrondobeitia *Facultad de Ciencias, Universidad del Pais Vasco, Appartado 644, 48080, Bilbao, Spain.*

K.J. Kirby *English Nature, Northminster House, Peterborough, PE1 1UA, United Kingdom.*

C. Lavers *Department of Geography, University of Nottingham, Nottingham, NG7 2RD, United Kingdom.*

N. Lust *Laboratory of Forestry, State University of Ghent, Geraardsbergse Steenweg 267, B 9090 Gontrode, Belgium.*

P. Di Martino *Viale Manzoni 59, 86100, Campobasso, Italy.*

P. Mills *Environmental Sciences Unit, Trinity College, Dublin 2, Ireland.*

C. Montanari *Istituto di Storia Moderna e Contemporanea, University of Genoa, Via Balbi 6, 16126, Genoa, Italy.*

D. Moreno *Istituto di Storia Moderna e Contemporanea, University of Genoa, Via Balbi 6, 16126, Genoa, Italy.*

B. Muys *Laboratory of Forestry, State University of Ghent, Geraardsbergse Steenweg 267, B 9090 Gontrode, Belgium.*

G.F. Peterken *Joint Nature Conservation Committee, Monkstone House, Peterborough, PE1 1UA, United Kingdom.*

P. Piussi *Istituto di Selvicoltura, Universita' di Firenze, Via San Bonaventura 13, 50145, Florence, Italy.*

F. Salbitano *Istituto di Selvicoltura, Universita' di Firenze, Via San Bonaventura 13, 50145, Florence, Italy.*

M. Saloña *Facultad de Ciencias, Universidad del Pais Vasco, Appartado 644, 48080, Bilbao, Spain.*

C. Watkins *Department of Geography, University of Nottingham, Nottingham, NG7 2RD, United Kingdom.*

A.G. Williams *Department of Geographical Sciences, University of Plymouth, Drake Circus, Plymouth, PL4 8AA, United Kingdom.*

Preface

The overproduction of food within the European Community (EC) means that there is a large area of agricultural land that is surplus to requirements. At the same time, there is a continued demand for wood and wood products. As a result of these two factors there is increasing interest in afforestation in Europe. This pressure for more woodland has come to the fore at a time when there is considerable concern about environmental change. Yet the ecological and environmental effects of afforestation are not well known. This is partly because the processes concerned are very complex and partly because the effects vary so much depending on where the afforestation takes place, what trees are involved and how they are managed.

The A1 Working Group (on the retrospective study of man-induced changes in European forest ecosystems) of the Forest Ecosystem Research Network (FERN), part of the European Science Foundation, organized a meeting at Freiburg in September 1991. The aim was to bring together researchers from different countries and disciplines to consider aspects of the history and ecology of European afforestation, both natural and artificial. The meeting stemmed from an earlier workshop (*Human Influence on Forest Ecosystems Development in Europe*), which was held at Trento in 1988 (Salbitano, 1988). The 3-day meeting was held concurrently with the meeting on the *History of Small-scale Peasant Forestry* organized by Professor Dr H. Brandl for the International Union of Forest Research Organizations (IUFRO). It was attended by 41 participants from 14 countries, namely: Bangladesh, Belgium, Canada, Czechoslovakia, Finland, Germany, Ireland, The Netherlands, Portugal, India, Italy, Spain, Sweden and the United Kingdom.

A key theme of the meeting was the importance of using a wide variety of sources and methods of approach when studying long-term ecological and land-use changes. It was recognized that approaches based on those developed by ecological historians were particularly valuable. It was also recognized that interdisciplinary studies were of great importance. One of the main conclusions of the meeting was the need to keep in place the strong interdisciplinary and international links that were brought about by the FERN A1 Working Group in order to facilitate further research.

I would like to thank Professor Dr H. Brandl, Professor P. Piussi, Professor C.O. Tamm, Professor H.W. Zottl, Dr M. Mahnig, Ms A. Teller, Mr R. Wirz, Mr J. Scham and Mrs J. Maclaran for their assistance in organizing the meeting. Dr R. Balzaretti, Dr B. Hart and Mr C. Lavers provided editorial advice. Mr C. Lewis and Ms Elaine Watts redrew many of the maps and diagrams.

Charles Watkins
Nottingham, June 1992

1

Forest Expansion and Nature Conservation

C. WATKINS

SUMMARY

This chapter considers some of the implications for nature conservation of the expansion of woodland with special reference to the situation in lowland Britain. The principal considerations that should be taken into account when planning the location and detailed siting of new woodland are outlined. The need to avoid establishing woodland on existing semi-natural sites is emphasized.

FOREST EXPANSION

There have now been several good historical studies that show how the area and distribution of woodland has fluctuated in many European countries over the last two centuries (Salbitano, 1988). In some areas, such as in parts of the karst landscape of Slovenia (Anko, 1988) or the Kielder Forest of Northumberland, England (Wallace *et al.* 1992), there has been massive afforestation; in other areas, such as parts of Molise, Italy (Chapter 6), there has been much clearance. Although there is considerable regional variation in forest gain and loss, in very broad terms the area of woodland has been increasing over this period.

Studies of woodland loss and gain at the local scale show that the detailed pattern of woodland change is very complex. The gross change in woodland distribution over a period of time is frequently greater than the net change. A detailed census of woodland change in Nottinghamshire, England, for the period 1920–1980 showed, for example, that there had been an overall increase of 27% (3535 ha) in the area of woodland. When gains (5482 ha) and losses (2320 ha) were taken into account, however, it was discovered that the gross change in woodland area was 2.3 times greater than the normally quoted net change arrived at by examination of official estimates of woodland area (Wheeler, 1984; 1988).

New areas of woodland can be established by planting, by natural regeneration or by a combination of these methods. Which method is used depends on a wide variety of social and economic factors. In many upland parts of Europe land abandonment and the reduction of grazing pressure has resulted in the natural regeneration of woodland (see Chapters 4, 5, and 7). In Britain, although there has been small-scale natural regeneration of woodland on old commonland, the principal form of woodland establishment is planting especially, until recently, of the uplands.

The area of woodland in Britain has doubled in the 20th century largely due to the government policy of woodland expansion, which has its roots in a series of late 19th-century government reports, but which was essentially established with the formation of the Forestry Commission in 1919. The long-term aim was to establish a suitable strategic reserve of timber. The target area of 2×10^6 ha has now been reached and in the 1980s there was a general questioning of the assumptions underlying previous forestry strategy (Mather, 1991). In addition to economic critiques (National Audit Office, 1986), the long-standing arguments that coniferous afforestation in the upland areas was causing significant problems in terms of reducing the nature conservation value of the planted land, reducing the quality and extent of public access on hill land and damaging the scenic quality of the uplands came to a head (Tompkins, 1989). One key issue was the battle over the afforestation of parts of the Flow country of Caithness and Sutherland (Stroud *et al.*, 1987; Lindsay *et al.*, 1988) which forms the background to Chapter 9.

General concern about agricultural surpluses in the EC has prompted consideration of forestry as an alternative land use. The need to withdraw land from the utilized agricultural area of the EC (Lee, 1991) has led to a policy shift away from planting in the English uplands (other than re-stocking), towards lowland afforestation, including schemes such as the MAFF Farm Woodland Schemes (Gasson and Hill, 1990), the new National Forest in the Midlands (Countryside Commission, 1989a) and a series of community forests around large towns and cities (Countryside Commission, 1989b). The shift in interest to lowland woodland management is also reflected in the Forestry Commission's broadleaved policy (Watkins, 1986).

In addition, there is greater interest in timber as a farm crop. This is derived from a desire to diversify the farm economic base and a concern for high-quality timber to offset the generally poor-quality timber associated with volume upland plantations. These aims could be achieved by developing farm-forestry and agroforestry systems (REEB, 1991).

Current estimates suggest that around 1.2×10^6 ha of arable land may be taken out of production (*The Times*, May 22, 1992). By December 1991, 154 579 ha had been set-aside under government schemes (MAFF, 1991). Various new uses for unwanted arable land have been proposed and introduced (Potter *et al.*, 1991). These include the setting aside of agricultural land as fallow, and the use of land for 'soft' non-agricultural uses, such as golf courses and country parks. The House of Commons Agriculture Committee's Report on Land Use and Forestry suggests that these types of land-use change are only likely to use up around half the surplus land available. They conclude that the 'significant alternative land-use in the next 20 years is likely to be forestry' and that 'planting targets since World War II have aimed at a modest expansion in the country's forestry estate: there now seems to be the scope, if not the necessity, for a far greater emphasis on the role of woodlands and forestry in the process of rural development' (House of Commons, 1990, p.xv).

The current surplus of agricultural land means that one of the principal locational factors affecting afforestation in Britain throughout the 20th century, namely that afforestation should only take place on 'unimproved' land of low agricultural value, is no longer of paramount importance. This change is welcomed by many conservationists, as upland areas contain the largest areas of semi-natural vegetation remaining in Britain; these areas are specially important from the point of view of nature conservation because they have not been fragmented by agricultural intensification as in most of lowland Britain.

The important question as to where the new woodland should be established is now raised. In the past, when significant changes in the agricultural economy resulted in land surplus, land was frequently abandoned and secondary woodland developed in a haphazard manner. This was the case in parts of Britain in the late 19th century when the agricultural depression of the time resulted in many farm bankruptcies. Examples of such unplanned land abandonment in Italy are discussed in detail in Chapters 4, 5 and 7. The current land surplus is rather different in that it provides the opportunity to plan the best location of the new woodland.

Different European government bodies are drawing up specifications as to which specific types of land should receive grant aid for woodland establishment, and which sites should preferably not be planted (see Chapter 13). This chapter considers the locational and site factors that should be considered in lowland Britain if significant nature conservation benefits are to be achieved by forest expansion.

THE LOCATION OF THE NEW WOODLAND

Until recently, there was no significant 'regional component' within British forest policy. During the 1980s, however, several commentators stressed the need for enhanced control, both over the allocation and design of future afforestation (Brotherton, 1986). The Countryside Commission called for planning controls, the Countryside Commission for Scotland (who experienced the greatest pressure for afforestation) argued only for planting licences, while the Convention of Scottish Local Authorities championed Indicative Forestry Strategies

(Selman, 1990). These have now been accepted at an official level in Scotland and on a more informal basis by some English and Welsh counties.

Forestry and woodland strategies are documents, usually drawn up by the local authority, which identify those areas that are suitable for different types of forestry and those areas that are best left unplanted. These strategies enable foresters to take account of sensitive areas when making planting proposals and identify areas where they are unlikely to encounter conflict. They could be a useful means of reducing disputes over the allocation of different land uses, and of bringing different interests together. However, to succeed they need 'to incorporate the full range' of land-use interests (House of Commons, 1990, p.xxii).

Another regional component has been introduced by the new National Forest and the various community forests, which have been scheduled to be established near large towns. Within these areas, landowners receive higher levels of grant aid when establishing new woodland – especially if they allow increased public access on their land. In practice, there may well be considerable difficulties in encouraging landowners to establish new areas of woodland. Indeed, one of the conclusions of a recent research project is that existing grant structures appear to be 'insufficient to implement the concept of Community Forests' (Bishop, 1990, p.402). This project included surveys of farmers, property developers and mineral companies. The results of these surveys indicated that, on farmland, planting was likely to be limited and would tend to be associated with 'hobby' farmers and on farms where game management was an important consideration. This study reinforces the results of other studies into the motives behind woodland planting and the relative unpopularity of various woodland grant schemes (Watkins, 1984; Gasson and Hill, 1990).

Property developers may carry out small-scale tree planting in advance of development in order to increase their chances of obtaining planning permission, or as part of a planning gain package – especially if the development was associated with leisure. However, all types of developer considered afforestation of their land holdings to be unrealistic. They were concerned about costs and problems of management and considered that woodland establishment would decrease the chance of obtaining planning permission in the future. A general conclusion that can be drawn from the results of this and other studies is that the levels of grant payable to landowners who establish new areas of woodland will probably have to be increased considerably if substantial areas of land are to be converted to woodland.

One general problem identified by the Bishop study is that the establishment of new areas of woodland is frequently made more difficult by the increasing use of new forms of tenure. Within agriculture there are now many types of short-term tenancies, and these are increasing the already diverse nature of land occupation (Ward *et al.*, 1990, Winter *et al.*, 1990). In addition, much land, although owner-occupied, is covered by options and conditional contracts with developers and mineral companies. Bishop considers that the effect of these legal factors is to 'increase land-use uncertainty' and restrict the 'chances for multi-purpose woodland creation . . .' (Bishop, 1990, p.409).

POTENTIAL BENEFITS OF LOWLAND AFFORESTATION

The national nature conservation body, English Nature, has been quick to see the potential nature conservation benefits of lowland afforestation (Watkins, 1991). They have argued that the afforestation of large areas of lowland agricultural land should, at the national scale, reduce the pressure to afforest land of nature conservation importance and that the new planting could contribute substantially to the national planting target of 33 000 ha per year. In practice, it appears likely that there will be continued pressure for upland afforestation. However, the increased scope for new commercial woodlands in the lowlands may reduce the pressure to manage ancient woodland in a manner unsympathetic to the nature conservation interest.

The establishment of woodland on arable and improved grassland will usually result in a net gain in terms of nature conservation. Coniferous plantations provide a better wildlife habitat than intensively farmed land. There will also be opportunities to create habitats of nature-conservation interest on derelict industrial sites.

There are several potential silvicultural benefits arising from the new lowland woodlands (Watkins, 1991). The new forests will be able to take advantage of a wide range of commercial species, both broadleaved and coniferous. In the uplands, the range is generally limited to Sitka spruce, lodgepole pine and the larches. In the lowlands, greater emphasis could be placed on Douglas fir, Corsican pine and broadleaved species. As well as a greater range of species, there is also the opportunity to increase the quality of timber coming onto the market.

In lowland forestry, a greater range of silvicultural systems is applicable than in the uplands, and the use of selection and shelterwood systems is more feasible. This will tend to result in higher costs, but thinning is likely to be more practicable and there is a greater likelihood that an earlier and more positive cash flow can be achieved. Although the recent great storms of 1987 and 1990 have demonstrated that woodlands in all parts of the UK are subject to storm damage, the risk of windthrow is in general considerably less in the lowlands than in the uplands.

The new lowland woods and the associated open land will also provide increased opportunities for woodland recreation. This may, in turn, reduce the recreational pressure on existing semi-natural habitats of high conservation value such as ancient woodland and lowland heaths. Moreover, greater access to woodland could provide ample opportunity to explore, develop and demonstrate the merits of the multiple use of woodlands and associated open land.

Detailed Siting of New Woodland

The basic aim of nature conservation is to retain the current richness and natural variation of wildlife communities (NCC, 1990). The effects of the establishment of woodland on the nature conservation value of a site depends essentially on the characteristics of the site concerned. In some cases there will be an increase in nature conservation value, for example, if woodland were established in fields that had for many years been managed intensively to grow cereals. In other cases there will be a decrease in nature conservation value, for example, if an area of grassland composed of a large number of species of herbs and grasses were planted with trees.

Unfortunately, the effects that the establishment of woodland on a site will have on its nature conservation value are frequently less clear than the two examples just provided might suggest. An important first step is to discover the current nature conservation value of any proposed woodland site. It will then be possible to assess whether the proposed woodland is likely to result in a net benefit or loss in terms of conservation.

What is the best way to approach the classification of existing types of vegetation in terms of the selection of sites suitable for the establishment of woodland? A useful and fairly pragmatic method is to distinguish between semi-natural vegetation, other good wildlife habitat and artificial habitat. In practice, the boundaries between these three types are often blurred. It is, however, a useful three-fold distinction to bear in mind when considering the location of new areas of woodland (NCC, 1990).

Semi-Natural Habitat

In Britain the main reservoir of wildlife lies in the remaining areas of semi-natural vegetation. This can be defined as those plant communities that consist of native species and which retain structural features that correspond to natural vegetation, but which owe their character to some degree to human intervention. Owing to the pervasive influence of modern agricultural and forestry practices, most semi-natural vegetation in lowland Britain has been fragmented and is now found scattered in small pieces between large areas of predominantly artificial habitat. In the lowlands the main types of semi-natural habitat include ancient woodland and hedges, heathland, unimproved grasslands, long-established scrub, peatlands, sand-dune grassland and heath, and also unpolluted and little modified freshwater, including ponds and lakes, streams and rivers.

The importance of retaining the remaining areas of semi-natural habitat is emphasized when the massive loss of this type of habitat since 1945 is recognized. Surveys indicate, for example, that 95% of lowland unimproved neutral grassland including herb-rich hay meadows, 40% of lowland acidic heaths and 20% of the hedgerow length that existed in 1949 have been lost since that date

(NCC, 1984). Although some of the remaining areas of semi-natural vegetation are protected by statute as National Nature Reserves and Sites of Special Scientific Interest, most are not protected in this way. Great care, therefore, has to be taken to ensure that no areas of semi-natural vegetation are inadvertently destroyed by afforestation.

OTHER GOOD WILDLIFE HABITATS

Although the most important habitats for nature conservation in Britain are usually semi-natural, many species are able to thrive in situations where the vegetation has been substantially modified. This type of habitat is widely distributed throughout Britain and supports very large populations of common species, as well as some rare species. In some intensively managed areas it is now the principal wildlife resource. The specific habitats concerned are very varied indeed and include grassland that has been improved in the past but which is now not managed intensively, early colonist vegetation on former industrial and urban sites and mineral workings, and many hedgerows and some plantations.

New areas of woodland should only be established on other good wildlife habitat if there is a clear nature conservation benefit. If small broadleaved woods were established on some of the grassland in this category, for instance, there would be an increase in habitat diversity. Any new woodland should preferably consist of native tree species. Care should be taken to ensure that any new woodland does not affect adversely adjoining sites of conservation interest.

ARTIFICIAL HABITATS

This category includes artificial communities, principally crops and grassland. These communities generally consist of a very small number of individual species. Most of lowland Britain falls into this category and it includes intensively managed arable land, grassland sown within the last 10 years, derelict industrial and urban sites that few species have colonized and recently worked and restored mineral workings. Most artificial habitats will be improved, in conservation terms, by the establishment of woodland, whether it be coniferous or broadleaved, native or exotic. It is on this type of habitat that any areas of intensive commercial afforestation should take place as these plantations will support more wildlife than the land-use they replace.

SPECIFIC FEATURES OF CONSERVATION INTEREST

Although there will be nature conservation benefits from the establishment of new woodland on most areas of artificial habitat, and some areas of 'good wildlife habitat', there are some exceptions and special cases. These include:

Interesting local flora

There are some sites within areas of artificial habitat that have a flora of conservation value. Some arable fields, for example, still support rare arable weeds and even fairly recently abandoned mineral workings or industrial sites may have gained local importance following the establishment of orchid colonies or an interesting weed flora.

Parkland and hedgerow trees

Many areas of parkland have had their nature conservation interest reduced in recent years by the application of fertilizers and herbicides. However, even though the ground flora may now be of little interest, the remaining old parkland trees frequently survive. These trees are important both for their conservation interest – many contain rare lichen or invertebrate species – and their cultural value. The extensive planting of new trees close to such old trees may well result in competition for light and water, and restrict the dispersal of invertebrates. The best approach is likely to be to retain the area as parkland and to assist the natural regeneration of new young parkland trees derived from local genetic stock. Old hedgerow trees and pollards should also be retained.

Bird populations

Some areas of artificial habitat may be of great value to bird populations. Some breeding populations of rare birds such as the stone curlew, cirl bunting and Montagu's harrier make use of the stubble of improved arable land for autumn feeding. Some moorland birds, including waders and blackgame, make use of intensively managed grasslands near to moorland for feeding, and some fields are used for feeding and roosting by concentrations of both birds of passage and wintering birds.

Adjoining land

New woodland may affect the conservation value of adjoining land and water of conservation interest. For example, new plantations of poplar or willow may directly reduce the conservation value of adjoining marshes or fens. Drainage water from new plantations may alter the hydrochemistry or sediment load of downstream rivers or lakes. Wider aspects of the effects of afforestation on water acidity are discussed in Chapter 11. These factors should be taken into consideration when locating new areas of woodland (Forestry Commission, 1988).

GENERAL SITING CONSIDERATIONS

Concepts drawn from cultural landscape history and landscape ecology may be used to inform the detailed planning of the woodland areas. Like most of Europe, Britain has a remarkably diverse landscape that has developed through the complicated interplay of natural characteristics and human activity. There is tremendous regional variation in such features of the landscape as the layout of field patterns, the proportion of woodland and the way it is distributed, the type of settlement pattern, the framework of lanes and rights of way, the area of commonland and so forth.

The detailed grain of the landscape has frequently been strongly influenced by historic legal and cultural factors such as the type of field systems, the nature of land ownership and the density of the population (Hoskins, 1955; Rackham, 1986). In addition, there are local variations that have resulted from particular agricultural and woodland management practices. Such variations include the number and species of hedgerow trees and whether they have been pollarded, the type of hedging that is practised, the distribution of old orchards and parks, and the width of road verges (see also Chapters 5, 6 and 7).

All these aspects of the cultural landscape are important in giving an area a particular sense of place. Local features of interest should be protected and every effort should be taken to ensure that the new areas of woodland complement the area's cultural landscape. Of course, the new woodland itself will be a novel element of the cultural landscape.

Landscape ecology consists of the study of the structure and function of the landscape and how it changes over time. It is concerned with the relationships between spatial patterns, such as the mosaic of different types of habitat, and ecological processes, such as the expansion and contraction of species populations (Selman, 1993). A recent review of the implications of landscape ecology for landscape planning concludes that there is considerable room for further detailed research, but identifies several landscape ecology principles (Selman and Doar, 1992). In terms of the establishment of new woodland, perhaps the most important aspects identified relate to the size and structure of woodland blocks, the proximity of woodland blocks to each other, and the way in which such blocks are connected to each other by hedges and other habitat links or wildlife corridors. Although there is still need for considerable research into the precise mechanisms behind the theory of landscape ecology (see Chapter 3), the following factors need to be taken into account when deciding the layout of new areas of woodland.

The size of individual woods

There has been much discussion about the effects of size on the nature conservation value of woodland (Helliwell, 1976). In very general terms, the larger the block of woodland, the better the potential conservation value. The interior of large woods is partially protected from agricultural pollution and ground water eutrophication and is generally less affected by adjoining land uses than that of small woods. Larger woods are likely to have a greater total number of species and larger populations of individual species. This is especially true of species that are characteristic of the forest interior and for the more uncommon woodland species. Large woods are also likely to have greater diversity of habitat than small woods.

In general, therefore, there is every reason, from the point of view of nature conservation, for establishing large new blocks of woodland in the new forests. However, some small woods, such as small areas of ancient semi-natural woodland are a very important conservation resource. There are many small areas of semi-natural scrub that support a wide variety of species. Moreover, small woods and narrow strips of woodland may play an important part in linking larger areas of woodland. Many farmers and landowners may wish to establish woodland partly for game conservation and, in this case, small woods with a large proportion of edge habitat will be beneficial for pheasants.

New broadleaved woodland will benefit from being established next to ancient woods. This will help to protect woodland species already within the core woodland, and may enable them gradually to spread into the new woodland. Certain species, such as bluebell and primrose, may colonize the new woodland fairly quickly, although some characteristic ancient woodland indicators, such as herb paris, are likely to take a very long time to spread (Rackham, 1980; Peterken and Game, 1984; see also Chapter 3). There will be cases when it will be inadvisable to enlarge woods in the interests of nature conservation. This would be so if, for example, such a wood were surrounded by semi-natural grassland.

Woodland shape

The woodland edge forms a valuable habitat for many bird and insect species and, in general, the greater the amount of edge habitat the better a wood is in terms of nature conservation. The simplest way to increase the amount of edge habitat is to scallop the edges of the woodland. In terms of the overall shape of the wood, a long thin wood has more edge habitat than a compact square or round wood. The main disadvantages of such thin woods are that a greater proportion of the wood can be prone to damage from spray drift and a smaller proportion of the wood is suitable for species characteristic of the woodland interior. The best plan is therefore to ensure that there is a mixture of different shapes of wood within the new forests.

Woodland density

The fact that around half of the area covered by the new forests will eventually be wooded will itself bring nature conservation benefits. In general, woods in well-wooded lowland landscapes contain a more diverse wildlife community than those in landscapes with little woodland. In addition, woods that are physically closer to other woods tend to be richer in bird species – especially those characteristic of the forest interior.

Wildlife corridors and woodland 'connectivity'

The linking of the woodland blocks by even simple hedgerows may serve a useful ecological function, although more research is needed on the value of such 'corridors' to different groups of species (Peterken, 1992). The movement of birds, small mammals and carabid beetles along the hedgerows can help to maintain local populations. More woodland plant and bird species are associated with tall hedges over 1.5 m high and wide hedges around 4 m in width than with simple hedges. Hedges also become more effective in their ecological function as their structure becomes more complicated. Hedges with ditches and hedgerow trees are of particular benefit to woodland carabid beetles and birds. If linking hedges are thin it might be advantageous to thicken them up in order to improve their value as wildlife corridors. Care should be taken to ensure that any thickening of hedges does not destroy existing grassland habitat of nature conservation value. Existing ancient hedgerows should be maintained, and they could be used to form the boundaries of new woodland blocks.

Woodland disposition

The way in which woodland is disposed in the landscape can influence the ecological value of an area. The 'patchiness' or the pattern of different types of habitat affects the degree of ecological interaction between the different habitats (see Chapter 2). If a landscape is very 'patchy' the abundance of species that require two or more habitat types to thrive will increase, but at the expense of species that prefer the interior of specific types of habitat. Some small-scale semi-natural habitats can be incorporated into new woods, but this will depend on both the requirements of the habitat and the way in which the new woodland is managed. For example, small fields of herb-rich grassland can be tied in with the ride network of a new area of woodland and be managed to maximize their nature conservation interest. The same might well be true of areas of scrub. On the other hand, new areas of woodland should generally be kept well away from small areas of heathland to reduce the threat of shading and natural regeneration of trees. Indeed, every effort should be made to expand the areas of habitat such as heathland and wet pastures rather than to establish woodland on them.

Woodland composition

The composition of woods in terms of both tree species and habitat type is obviously of crucial importance. In lowland Britain, woods consisting primarily of broadleaved trees, with a relatively small proportion of conifers, tend to be richer in wildlife than those consisting largely of conifers. Woods in which there is a mixture of different types of habitat suitable for both woodland edge and woodland interior species are likely to maximize the abundance of wildlife.

NEW FOREST MAPS

All the existing information about land use and habitats within the proposed new forest areas should be collated to form a map of the area of the new forest. This map will show the principal factors that have to be taken into account when determining the location of the areas where woodland is to be established. It is likely to show the following features:

- the existing semi-natural habitat;
- the existing good wildlife habitat;
- areas of artificial habitat;
- specific features of conservation interest;
- existing and planned infrastructure including railways, motorways, roads, public bridleways and public footpaths, pipelines and overhead electricity power lines;
- existing and planned industrial areas and residential areas;
- existing and planned quarries, mines and gravel extraction;
- existing and planned recreational sites;
- existing sites of archaeological or historic importance.

Geographical information systems could be used to predict the mosaic of different types of habitat and land-uses within the new forest areas. Such an analysis could in Britain be based, for example, on the Institute of Terrestrial Ecology's land classification. Once the prediction had been made, it would be possible to generate a variety of proposals concerning the location of the new woods. Further analysis could be carried out to indicate the potential for new habitat creation and the management requirements of the new woodland. It might also be possible to recognize zones with differing requirements within the forest. The system might be developed as a medium-term monitoring device to check on the types of different habitat that have been afforested (Haines-Young and Ward, 1991).

Once the initial mapping has been completed, it should be possible to identify the broad areas of land suitable for different types of forestry. For example, the areas of artificial habitat that are not likely to be affected by other forms of development could well be suitable for commercial timber production. Sites where new woodland would provide nature conservation and recreational benefits could also be identified. In addition, and very importantly from the nature conservation point of view, the map will show areas of semi-natural habitat where no new woodland should be established.

The proportion of the designated forests that will be wooded will depend on a wide range of local factors. In general, it is expected that the total area of woodland in the Community Forests would not exceed 55% of the total land area (Countryside Commission, 1991). Such proportions occur at present in relatively few locations such as the Forest of Dean, Sherwood Forest and parts of the Breckland. Ultimately, of course, the decision whether or not to establish new areas of woodland will rest with the landowner. Any forest plan must therefore be fairly broad, general and indicative – particularly as regards land allocation.

CONCLUSIONS

The surplus of agricultural land in Europe provides an opportunity to increase substantially the area of woodland. If this expansion is to result in significant nature conservation benefits, it will have to be planned very carefully. This is especially true for those parts of Europe, such as lowland Britain, that have lost most of their semi-natural habitat in the present century.

There is now considerable public support for the establishment of new woodland in the interests of nature conservation. The idea of a new national forest in lowland England has received much publicity. A similar new forest is being established in the Scottish lowlands. There are also several local initiatives: a recent newspaper article has described the efforts of the small Findhorn Foundation, which 'aims to reconstruct the vast ancient Caledonian Forest that once covered much of the Scottish uplands with native Scots pines, and is now reduced to just 1 per cent of its original size' (*The Independent*, 2 May 1992, p.41). This small organization has attracted the support of the Forestry Commission and Scottish Natural Heritage and 105 ha of land has been fenced against deer in order that trees can be established.

There is room for much interdisciplinary and international research into the best way of encouraging landowners to establish woodland that will provide nature conservation benefits as well as a commercial return. The relative benefits of making use of natural regeneration, as opposed to establishing woodland by planting trees, need to be assessed on a regional basis. This chapter has stressed the need to take account of the existing habitat types, and the variations in cultural landscape within which the new woodlands are to be established. Further research is needed to increase our knowledge of the way the landscape has developed, and in order to establish where existing habitats of conservation importance are found. Further research is required into the best way of reconstructing habitats that have been lost (Buckley, 1989; Francis *et al.*, 1992). Geographical Information Systems may be a useful tool that can be used to determine the rate of loss of different types of habitat and to monitor the expansion of woodland.

2

The Effects of Plantation Management on Wildlife in Great Britain: Lessons from Ancient Woodland for the Development of Afforestation Sites

K.J. KIRBY

SUMMARY

Seventy-five percent of British forests are plantations and their extent is likely to continue to increase. The majority have been planted on open ground (afforestation) and are only now approaching the end of their first rotation. About 10% of British woodland is, however, plantation on ancient woodland sites in which a wider range of tree crops, rotation lengths, thinning regimes and coupe sizes can be found than in most new afforestation. The experience of how wildlife on ancient woodland sites has responded under plantations may be used to develop models for the treatment of afforestation sites in future to improve their value for nature conservation.

INTRODUCTION

Forests in Great Britain cover only 10% of the land surface, and about two-thirds of this consists of plantings on open land (afforestation) (Fig. 2.1). Afforestation has changed the appearance of many parts of the countryside and has had profound effects on wildlife over large areas of the uplands (NCC, 1986). Most of this new planting has taken place in the last 100 years and much in the last 50 years, so many of the woods are still young. Most of the descriptions of the wildlife they contain (see for example Hill, 1979; 1986; Good, 1987; Good *et al.*, 1990; Avery and Leslie, 1990) concentrate therefore on the plant and animal communities of the first rotation. However, these communities are likely to change in subsequent rotations because of varying management or changes in the nature of the forest itself.

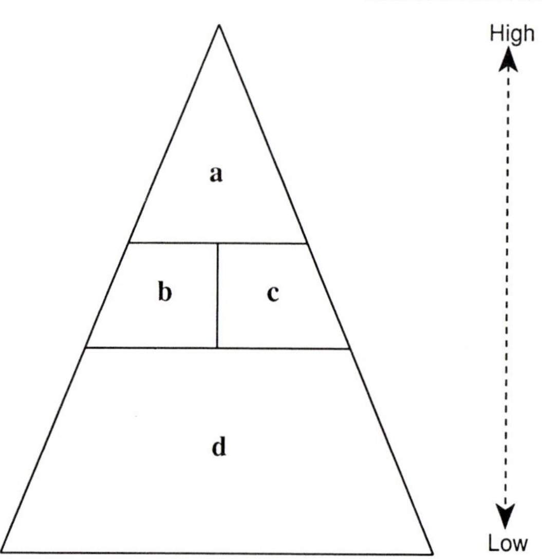

Fig. 2.1. The relative area of different woodland types in Great Britain. (a) = ancient semi-natural woods (15%): site wooded since 1600 AD, current composition consists of trees and shrubs native to the site that have arisen by natural regeneration, from stump regrowth (as in coppice) or from pollard regrowth; (b) = ancient sites now bearing plantations (10%): site wooded since 1600 AD; current tree cover has been planted and is usually of non-native species; (c) = recent semi-natural woods (10%): site has changed from open ground to woodland since 1600 AD through natural colonization by native trees and shrubs; and (d) = recent plantations (65%): site has changed from open ground to woodland since 1600 AD through planting, usually of non-native species.

Cunningham (1991) has assembled the arguments for continuing to expand the area under forests (largely through new planting) but there is now more emphasis on planting on lowland sites to reduce agricultural surpluses and to improve the landscapes around towns and cities. There is much more potential in such plantings to create rich and varied wildlife habitats than in past afforestation (Watkins, 1991), and nature conservation will often be as important an objective of management as timber production.

Models are needed therefore of how the wildlife in plantations behaves, to guide both the restructuring of existing afforestation sites and the development of new plantations in the lowlands. One input to such models might be data from studies of plantations within existing woods, which encompass more lowland sites, richer mineral soils and a wider range of tree crops, rotation lengths, thinning regimes and coupe sizes than can be found in most new afforestation.

Ancient Woodland and Plantations on Ancient Woodland Sites

On only about 2–3% of Britain has woodland cover existed continuously since before 1600 AD, and this woodland, regardless of its current composition or structure, is defined as 'ancient' (see Fig. 2.1) (Peterken, 1977a). It has been identified on a county-by-county basis from studies of old maps, aerial photographs and existing botanical and forestry surveys by the Nature Conservancy Council (Roberts *et al.*, 1992; Spencer and Kirby, 1992). The importance of ancient woodland for nature conservation derives partly from the species that are preferentially associated with it, which rarely occur in woods of recent origin (see for example Chapter 3 and Peterken, 1974). However, many woodland species are not so restricted in their distribution, so some extrapolation about the ecology of plantations from ancient to recent sites is possible.

In England and Wales, 40% of existing ancient woodland has been replanted (Spencer and Kirby, 1992), with the loss of much of its nature conservation value; semi-natural tree and shrub communities are destroyed and the richness of the ground flora is frequently reduced (Kirby, 1988). However, in some plantations groups such as butterflies have flourished (Peachey, 1980), whereas they have declined on nature reserves (Pollard, 1982). Understanding how and why such differences exist may help us in efforts to restore plantations on ancient sites to a more natural state (Kirby and May, 1989), to devise forestry regimes that allow the special characteristics of ancient semi-natural woods to survive (Forestry Commission, 1985; 1989) and to provide guidance for the design and management of new plantings on open ground.

What Sorts of Change Occur within Plantations?

The planting of large areas of open ground leads to the replacement of open-ground communities with those more typical of woodland. Within an existing wood there may also be changes following planting arising from the effects of a new tree crop species (especially the conversion of broadleaved woods to conifers). The effects of different crop trees on the ground flora have been well-studied (see, for example, Ovington, 1955; Anderson, 1979; Kirby, 1988; Pigott, 1990; Simmons and Buckley, 1992). They are not considered further here, because the choice of main-crop tree is generally limited in plantations by factors such as soil type, exposure or the market for the timber.

The creation of a plantation in a wood may also alter its structure, for example the number of layers in a stand and the disposition of stands of different ages. Wildlife may also be affected by operations such as drainage, fertilization and the use of herbicides designed to ensure that the planted trees survive (Mitchell and Kirby, 1989; Hunter, 1990).

Such changes must be related to the variation in environmental conditions and hence in the occurrence or abundance of many species through a crop rotation. Whereas the plant and animal communities of open ground may be relatively stable from year to year (under a constant external environment), progressive or cyclic changes within a forest are common. Thus Page (1968) and Hallbacken and Tamm (1986) point to the regular changes in soil pH that occur as a tree crop matures, but these changes may become apparent only over a long period. The interactions between changes in species abundance through a rotation and the structure of the forest are explored in the rest of this chapter.

Change through the rotation

LONG-ESTABLISHED WOODS

Species richness for higher plants is usually high when a stand is young and has an open canopy, and declines as the trees close canopy and form a thicket, but may then increase gradually as the trees are thinned and more light reaches the forest floor (Fig. 2.2a; Hill, 1986; Kirby, 1988). Annuals and biennials, such as *Poa annua*, *Senecio aquaticus* and *Sonchus oleraceus*, many of which have a ruderal life-strategy (Grime *et al.*, 1988), increase because of the opportunities for establishment provided by the bare ground created during felling and extraction. Relative to the unfelled stand, increased light, available nutrients and moisture (usually) then favour the growth of competitive species such as *Cirsium arvense* and *Chamaenerion angustifolium* (Fig. 2.2b; Grime, 1987). An alternative way of describing the changes is through the use of Ellenberg's indicator values for light, moisture and nutrients (Ellenberg, 1988): species found in clear-fells show higher index values than those in the unfelled stands. The spread of competitive species may reduce species density from its post-felling high even before the new tree crop is established sufficiently to influence the ground flora (Kirby, 1990). Once the thicket has formed however, plants that are tolerant of the conditions found under mature trees (low light levels and high root competition), such as *Anemone nemorosa*, *Carex sylvatica* and *Luzula pilosa* are relatively favoured and so likely to spread.

Cyclical changes in species abundance occur for other groups of organisms but are not necessarily in phase with the changes in ground flora richness. Many warblers are most abundant at the thicket stage, when the ground flora is least varied, whereas thrushes become commoner in more mature stands. Hole-nesting birds such as woodpeckers may breed only in the oldest stands (Fuller and Moreton, 1987; Smith *et al.*, 1987).

EVEN-AGED STANDS IN NEW AFFORESTATION

The general pattern of change in plant richness through a rotation in existing woods is likely to apply in new afforested stands (Hill, 1986) but there are additional aspects to consider – particularly how far the forest manager may be

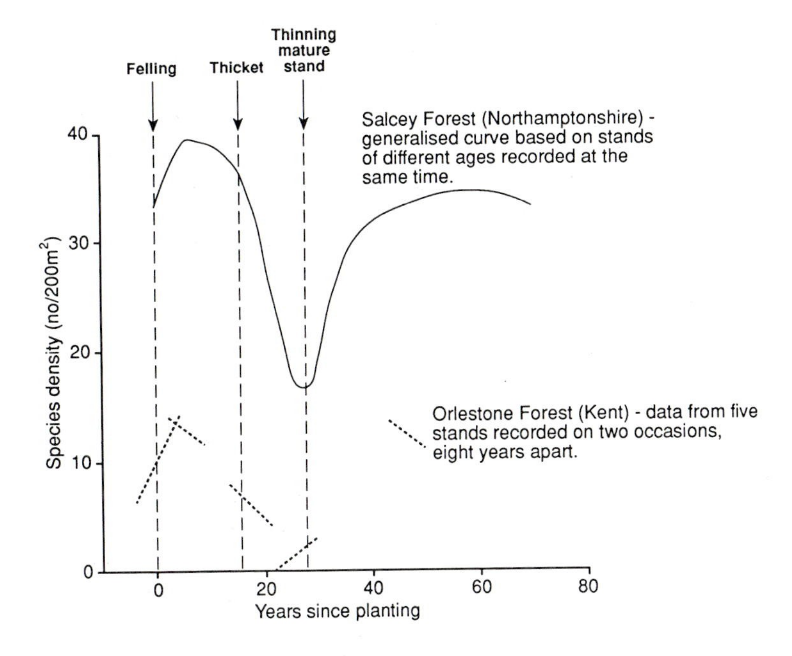

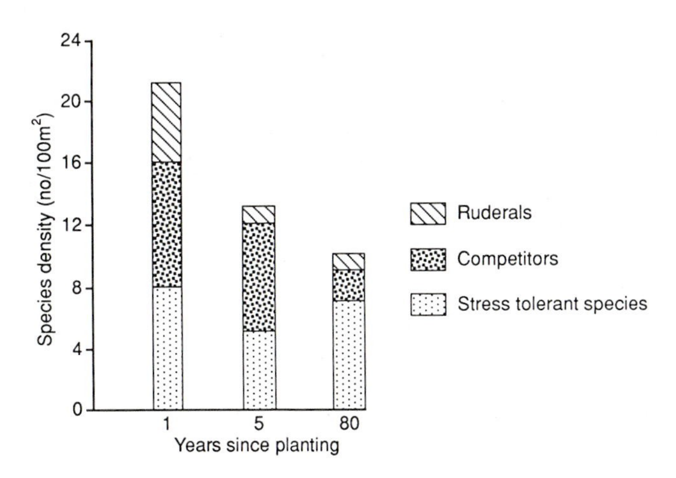

Fig. 2.2. Changes in: (above) species density of the ground flora (number per 200 m^2 quadrat) through a rotation, generalized from Kirby (1988) and unpublished data; and (below) the proportions of ruderal, competitive and stress-tolerant species (Grime *et al.*, 1988) after felling compared to old stands.

able to alter the treatment of the stand to benefit wildlife. There is some surface soil disturbance associated with new planting (usually ploughing) although this is less extensive but deeper than after felling an existing stand. Hence an increase in ruderal species is likely. A rapid increase in competitive species is also common because work is done (e.g. fertilization, drainage) to ameliorate the site or to remove grazing and prevent fires, which often previously limited plant growth, and incidentally promoted diversity. Thus areas of grassland left within plantations (rides and old meadows) may become overgrown and dominated by one or two rank grasses such as *Deschampsia cespitosa* (Good *et al.*, 1990).

In the planted areas, most such competitive species are eventually shaded-out as the canopy closes and the thicket forms. However, there are likely to be few species from the original open ground community that are capable of surviving under a forest cover, and so the relative drop in species richness at the thicket stage is greater than in a long-established woodland. The invasion of the new forest by specialist woodland plants will be limited to those with efficient long-distance seed-dispersal mechanisms such as bramble, *Rubus fruticosus*, except in sites close to existing reservoirs of woodland species such as hedges and wooded streamsides (see Chapter 3).

Development of a more varied ground flora in the latter part of a rotation may be enhanced in some stands by heavier and earlier thinning or by extending the rotation length. The latter might be useful where the species invasion is limited by colonization – there will be more time for species to arrive by chance rare events. The thinning option benefits species that can colonize quickly, but which are limited by shade and litter levels in unthinned stands. In many upland plantations, the risks of windthrow prevent the forest manager adopting either approach, and even where they might be feasible they result in extra costs or delayed revenue. In stands where both are possible, the choice as to which is most cost-effective in wildlife terms should be influenced by the shape of the 'invasion curve' (Fig. 2.3). More work is needed on how this varies from site to site and for different groups of species, but there are likely to be advantages in developing both options.

In long-established woodland the cycles of species change through a rotation are likely to follow broadly similar patterns through successive rotations (Fig. 2.4a). This assumption is the basis for advocating maintenance of 'traditional management' such as coppicing in ancient woods (Peterken, 1977b). On newly afforested sites, several rotations may be needed before such relative stability is attained.

Species characteristic of the former open vegetation may not be totally eliminated in the first rotation, but the former vegetation is likely to occur at increasingly lower abundances (Fig. 2.4b) as cumulative changes occur in the site conditions (Hill, 1986). For example, the soils in the new forest may become more podsolized, better drained or more nutrient-rich, depending on the site, the crop and its treatment (Pyatt and Craven, 1979; Miles, 1986; Pyatt, 1990). Even if recovery at the end of each rotation is to 75% of the starting level for species

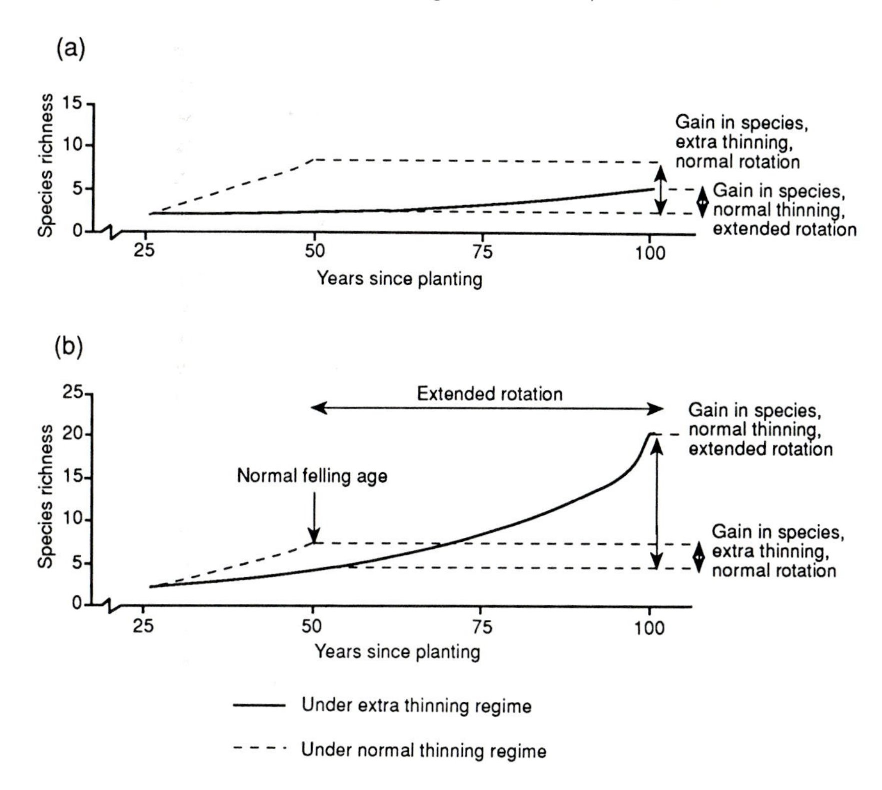

Fig. 2.3. Does heavy thinning or an extended rotation bring more benefits in terms of increased species-richness? A given level of income from timber production may be foregone for wildlife benefits in a stand either by increasing the extra thinning or by extending the rotation from 50 to 100 years. In site (a) it is most cost-effective for the manager to adopt extra thinning, in site (b), because of the different pattern of species invasion in the latter part of the rotation under normal thinning, it would be better to extend the rotation.

abundance, by the end of the fourth rotation the abundance has been reduced to less than half that at the time of the original planting. The same pattern of progressive loss over successive rotations, but of specialist woodland plants, may occur in long-established woods when light-canopied broadleaves are replaced by dense-canopied conifers or beech (Mitchell and Kirby, 1989).

Equally, there should be an increase with successive rotations in the abundance of woodland species (Fig. 2.4c). For highly mobile species, such as birds, the increase in richness over time may be very rapid but for ancient woodland plants the rate may be very slow – particularly in isolated woods (Peterken and Game, 1984; Chapter 3).

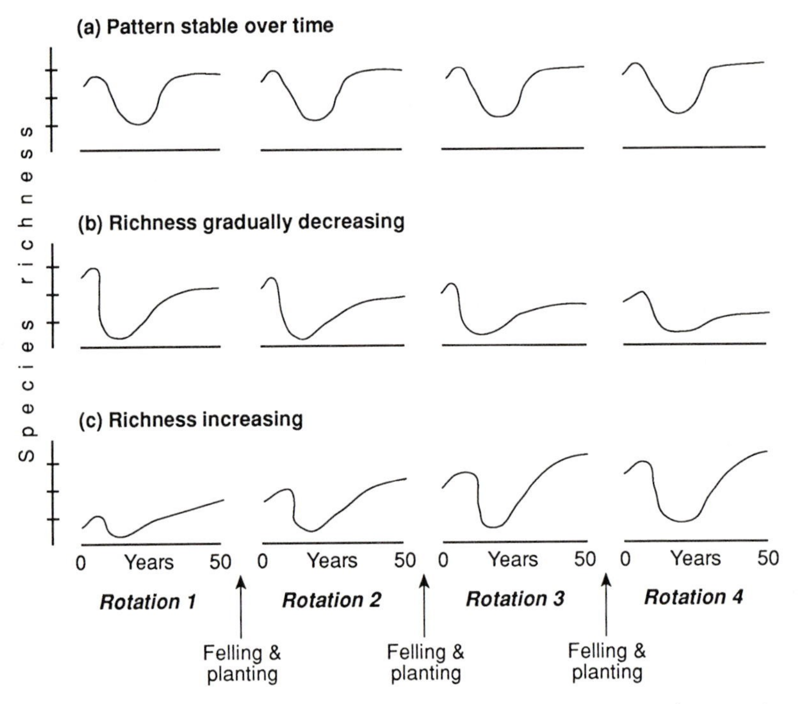

Fig. 2.4. Progressive change in species richness over four rotations for: (a) plants in a long-established woodland; (b) species of open ground on an afforestation site; and (c) species which invade the new forest.

Changes over time in a whole forest

Changes within a single even-aged stand may approximate to the pattern for the whole wood in many afforestation schemes where large areas are planted over a very short period. In the second and subsequent rotations differences in age-class between stands are likely to increase, either in a planned way, as some stands are felled early while others are held back, or through unplanned events such as windthrow.

The change (or lack of it) in diversity for the forest as a whole over time therefore comprises the combination of the changes for each individual stand (Fig. 2.5). The overall richness of a group of stands of different ages is likely to be as high or higher and to show less variation with time than a single-aged stand. Moss (1979), working in plantations of different ages, found 17 species of songbird in total, although the numbers for individual stands ranged from two to 14. In Orlestone Forest (Table 2.1) the number of plant species found in six stands combined was little different in 1991 from that recorded in 1983, even though some individual stands had increased and others decreased in richness over the period.

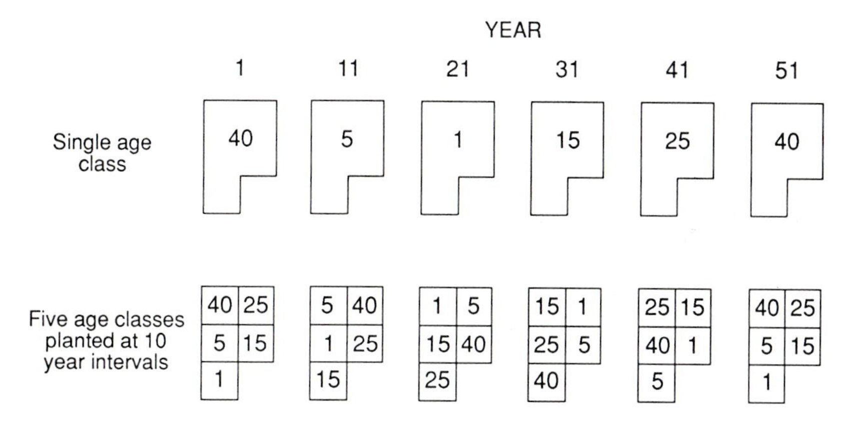

Fig. 2.5. Changes in species density (for a site similar to the upper curve in Fig. 2.2) for a wood composed of only one age-class compared with the same wood split into five evenly spaced age-classes. Felling takes place at Year 0 and Year 50. (Number of species is represented in boxes.)

Table 2.1. Number of vascular plants recorded in the ground flora in Orlestone Forest (Kent) in 1983 and 1991.

	No of species in five 200 m² quadrats per stand					
Stand	a	b	c	d	e	f
1983	25	14	15	19	2	8
1991	24	7	26	14	11	25
No of species in all stands combined						
1983	34					
1991	36					

A forest where all age-classes are present in equal proportions ('normal' in the forestry sense) may provide the most stable conditions for wildlife, but chronological age is not the best measure for judging normality, from a wildlife point of view. For much of the life of a stand minor differences in tree age (5 years or less for stand ages less than 100 years) have little effect on the environment created within the stand. Instead, four main growth stages may be distinguished:

- open-canopy stage;
- thicket;
- pole to young mature stands; and
- over-mature or veteran stands (Mitchell and Kirby, 1989).

The 'normal' forest to be aimed for might then be one where equal proportions of all these stages were present all of the time. Of these stages, the open-canopy stage and the veteran stand have the most distinctive wildlife communities, so in practice emphasis should be placed on increasing or at least maintaining these (Peterken, 1987).

Even with a defined age structure there are various options. How big should the individual stands be that make up a particular age band? How should these stands be distributed through a wood? And what should be done in forests where the age structure is already so skewed that normality could not be even approached for several rotations? How may a range of age-classes be maintained in practice in the many small plantations that exist on farms? The numbers of these small plantations are likely to increase under various European schemes designed partly to offset agricultural over-production (see Chapters 1 and 13).

Scale and arrangement of stands

In one British study, little difference was found between the development of the ground flora following felling in group fells of about 0.1–0.2 ha and that in 2–3 ha clear fells (Fig. 2.6). Coupe size, however, affects the response of other groups and species. The effect of clearing size on the regeneration of shade-tolerant *versus* less-tolerant species was, for example, described by Watt (1925) and has been particularly well documented in woods in North America (for example Leak and Filip, 1977) with small gaps favouring shade-tolerant trees.

Dormice, *Muscardinus avellanarius*, are reluctant to cross large clearings, so group fells dispersed through a wood would cause less disruption to their territories than the same area of clear fell (Bright and Morris, 1990). Such group fells might, however, be too small for butterflies of the open-canopy stage, and in the Blean Woods (Kent) fells of 1–3 ha were created to boost populations of the heath fritillary *Mellicta athalia* (Warren, 1987). In plantations on former grass heath, the woodlark *Lullula arborea* breeds in clear fells with large areas of short vegetation (Bowden and Hoblyn, 1990), which would not be provided by very small coupes.

The specialist birds found in Fenno-Scandinavia and North America which are declining following fragmentation of old-growth forests (Ripple *et al.*, 1991; Virkkala, 1991) are unlikely ever to be accommodated within plantation systems. However, there are smaller organisms (e.g. fungi, beetles, mosses) that are characteristic of old, undisturbed stands in Britain which could, in theory, survive in quite small areas. Are they favoured by small- or large-scale patterns? If some ground-living beetles are unwilling to cross roads, even forest roads (Mader, 1984), a 2 ha clear fell could represent a substantial barrier to movement.

Relatively little work has been done in Britain on optimum stand size for species other than those of the open-canopy stage, so the requirements of the latter are likely to dominate in much forest-design work. However, one should be

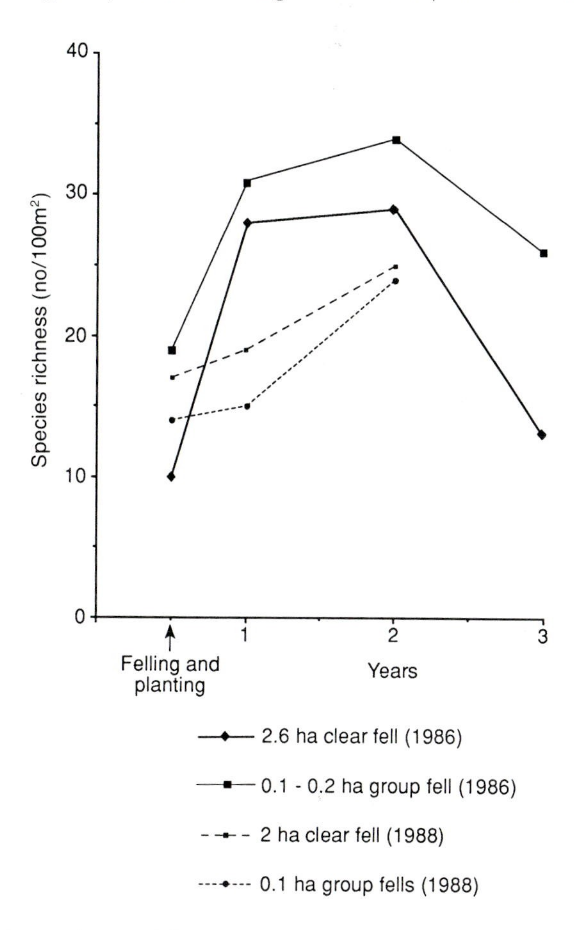

Fig. 2.6. Similarity of ground flora response in group and clear fells (Kirby, 1990).

aware that the patterns so created may not be the optimum for species that rely on the later growth stages and should be prepared to modify our recommendations accordingly if such situations arise. Circumstances that will rapidly improve conditions for a group of species in the short-term should also be distinguished from those that may provide a smaller immediate benefit, but one that is sustainable throughout the rotation. Felling 2 ha/year may provide superb conditions for butterflies, but in a 50 ha wood such a regime can be maintained for only 25 years, after which no new open space can be created in this way until the first stands have reached maturity, perhaps 25–75 years ahead.

Varying the distribution of stands of different ages also affects the species that use the edges between stands (Fuller and Warren, 1991) or make use of more than one age-class. Many birds of prey such as sparrowhawks, *Accipiter nisus*, may nest in old trees but hunt across more open ground such as rides or clear fells. Deer feed in open-stage stands but lie up in thickets. The restructuring of some upland forests to improve their design (Mackintosh, 1990), which will involve smaller compartments and more mixture of age-classes than in the first rotation therefore, has a cost in potentially increased deer damage. Capercaillie, *Tetrao urogallus*, similarly prefer a mixture of old and young growth (Leclercq, 1987), but the scale of the preferred mosaic is different and may be different again for other groups and species (Hunter, 1990).

The same total area of a given age-class may thus be distributed either as a few large stands, to minimize edges, or divided between many small stands to increase the length of edge. Stands in successive fellings may be grouped to help species restricted to one age-class spread from one stand to the next as they become suitable (which tends to reduce edge effects), or they may be dispersed, which benefits species that use stands of widely differing age-classes (Harris, 1984) (Fig. 2.7).

Coping with unbalanced age-classes and small woods

The preferred even distribution of age-classes may be difficult or impossible to achieve either because of the size of the wood or because of the age structure inherited from past plantings. Suitable designs for small woods may become more important in the future, if there is an increase in planting on lowland farms.

A coppice regime in which 0.5 ha is cut each year on a 20-year rotation could maintain continuity of open-stage conditions for butterflies or nightingales,

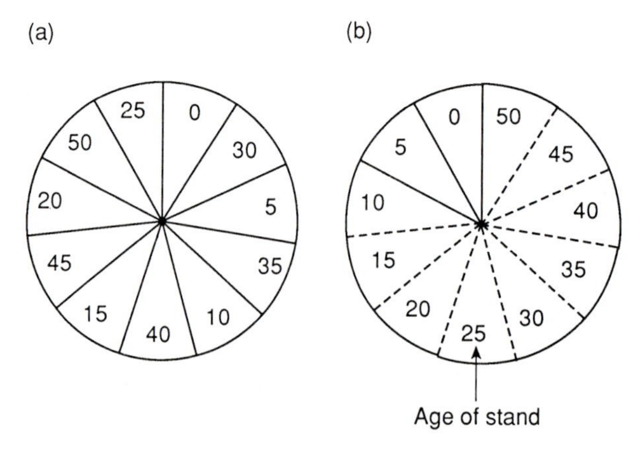

Fig. 2.7. Two arrangements of stands of different ages: (a) based on Harris (1984) designed to maximize the edge effect by dispersing stands; and (b) based on principles proposed by Warren and Fuller (1990) to ease spread of species and minimize edges. An 'edge' effect, shown by a continuous rather than dotted line between two stands is assumed to exist only if they differ by 10 years or more, or by 2 years or more in the first 10 years of a crop's life.

Luscinia megarhyncos (Bayes and Henderson, 1988) indefinitely in a 10 ha wood. To achieve the same degree of continuity and extent of open space (0.5 ha cut each year) in a wood managed as high forest on a 50-year rotation would require 25 ha, or 50 ha if the rotation were 100 years. If the wood were smaller than this, the annual cut of (say) 0.2 ha from a 10 ha wood on a 50-year rotation might be too small to maintain these species if the cuts were spread around the wood as in Figure 2.7a. Two options may then be considered. The first is still to have an annual cut, but group successive cuts as in Figure 2.7b so that the combined open area available at any one time is at least the desired 0.5 ha. Alternatively, 0.5 ha might still be cut at a time, but only every 2–3 years. Both options assume that the open-stage habitat remains suitable for the species concerned for several years after the felling and not just for the first year. This is usually the case for species of clear fells such as the woodlark in Thetford Forest (Bowden and Hoblyn, 1990).

In woods below 5 ha, however, even with the above measures it may not be practical to maintain a balanced age structure. The boom in planting that took place between 1945 and 1975 means that many larger forests also have stand age distributions very far from the normal. Thus Bernwood on the Oxfordshire/ Buckinghamshire border, was noted for its butterflies in the 19th and early 20th centuries when it was still a worked coppice (Thomas, 1987). It was felled and replanted almost completely over a 40-year period and the open stands so created boosted the populations of many of these species (Peachey, 1980; Thomas, 1988). Since 1970, however, relatively few open areas have been created by felling, and, if most of the crops are on a 50-year rotation, few more can be expected until 2005 with the next peak arriving between 2010 and 2020 (Fig. 2.8).

Plants typical of the open stage after felling may survive this unfavourable period as buried seed, although few of those found in coppices may survive in this way for more than about 50 years (Brown and Oosterhuis, 1981; Brown and Warr, 1992). There is no such option for the butterflies. Parts of the forest need to be managed specifically as refuges for these species to carry them through the period when few suitable areas may be created by normal felling operations. In Bernwood, and elsewhere (Warren and Fuller, 1990; Ferris-Kaan 1991), wide rides with scrubby edges (Fig. 2.9) have been created and maintained for this purpose.

Different sorts of refuges need to be designed into plantations for different groups of species. Areas of minimum intervention and trees left to grow old and die naturally are required as nesting sites for birds, and for dead-wood beetles and fungi, and they should also be encouraged in former coppice woods (Kirby, 1992). Refuges, if suitably managed, may also help to maintain some of the species and communities from the former land use, for example the old meadow areas within the Kielder Forest (Good *et al.*, 1990) or the populations of sand lizards, *Lacerta agilis*, surviving on heather within some of the Dorset heath plantations (Spellerberg, 1988).

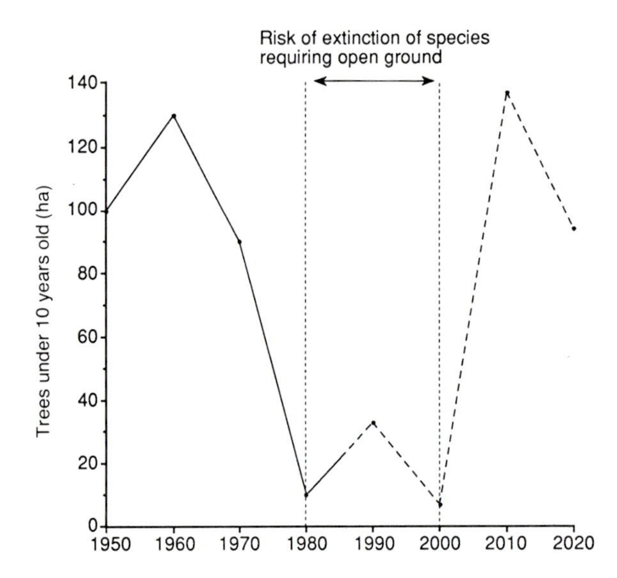

Fig. 2.8. Change in the extent of open stands (less than 10 years old) in part of Bernwood Forest, 1949–2020, assuming a 50-year rotation for most crops.

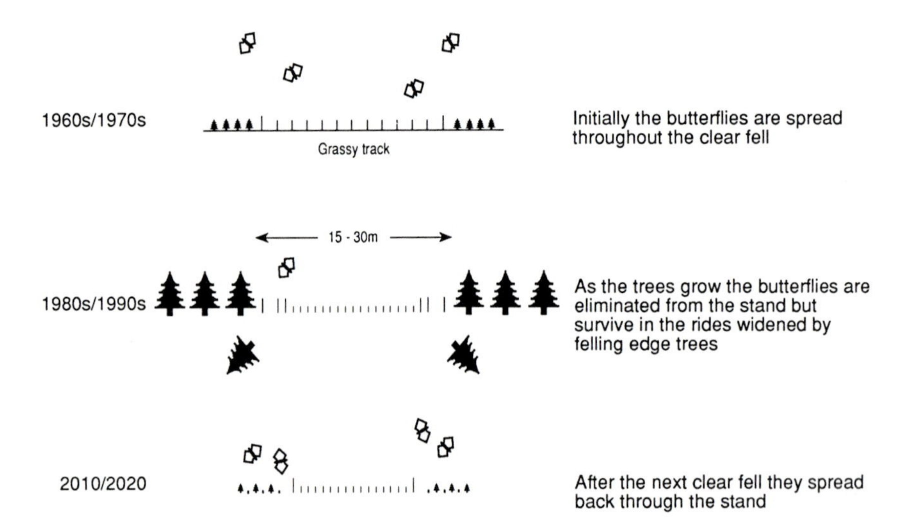

Fig. 2.9. Refuges created for butterflies along rides by felling edge trees.

The creation of refuges for groups of species is an attracti
foresters because it involves relatively little disruption to nor
forestry operations over the majority of the site. However refug
ever, maintain the full range of species that are associated with
in the forest cycle, and so must be clearly aimed at those that [illegible]
them. There is also always the risk that the commitment to such refuges may be lost under a change of policy or of staff, and if that occurs the species are very vulnerable to extinction because they are mostly concentrated into very small areas.

CONCLUSIONS

Not all the plants and animals found in plantations on ancient woodland sites are likely to occur in new afforestation, but there is some overlap and similar principles determine how they distribute themselves in time and space through a rotation. Cyclical changes in abundance of different species occur in individual stands. Those that are rich now are unlikely to remain so. Changes in abundance in individual stands must be integrated for the age structure of the wood as a whole to predict the future species richness of the site.

The scale and arrangements of stands should be varied through the plantations to benefit different groups of species. Once a pattern has been chosen, it should be maintained, to increase the likelihood that slow colonizers will be able to survive within the pattern. Where the age structure is unbalanced (and particularly in small woods) there is a need to maintain long-term refuges for the species of open-stage and over-mature stands.

Although the rate of new planting has declined since 1989, Government policy in the UK is still to increase home-grown timber production by expanding the area of new plantations. Such forests should be as diverse as possible so that they make a positive, albeit very distinct contribution to wildlife conservation in the 21st century. Learning from the changes that have occurred in plantations on ancient sites can help us to do this.

ACKNOWLEDGEMENTS

Richard Ferris-Kaan, Stefa Kaznowska, Peter Mitchell, George Peterken, Peter Savill and Rob Soutar kindly commented on the paper, while Caroline Bowie drew the figures.

3

Long-term Floristic Development of Woodland on Former Agricultural Land in Lincolnshire, England

G.F. PETERKEN

SUMMARY

The secondary woodlands that developed on former farmland in central Lincolnshire after 1600 now contain fewer vascular plant species than ancient woods in the same district. About 40% of the vascular plant species in ancient woods have been unable to colonize new woodland even after 400 years, and there is no sign that they will do so eventually. While renewed attempts to plant more woodland on farmland are welcome because they create additional wildlife habitat, most of the characteristic and attractive plants of ancient woodland will not colonize without help.

INTRODUCTION

Today much is heard about set-aside and tree-planting on agricultural land (e.g. Insley, 1988). It is claimed that wildlife will benefit because species that were hitherto excluded from cultivated ground will colonize the new woodlands. In the long-term it is supposed that these new woods will steadily acquire more species until they are as rich as those woodlands inherited from the original forests. In England these assumptions can be tested by examining the wildlife in secondary woodlands, which have developed on former agricultural land over the last 400 years. I shall first outline the results from an examination of the vascular plants in secondary woods in central Lincolnshire, which developed since 1600, then review briefly the evidence from secondary woods elsewhere.

THE WOODS

Central Lincolnshire was originally covered in woodland, save perhaps for the fens beside the river Witham. Most was cleared more than 2000 years ago and, by 1086, the surviving woodland was reduced to a scatter of coppices and wood pastures (Foster and Longley, 1924). These fragments of the original woodland were further cleared in historic times until they were reduced to just 89 separate fragments covering a total of 2543 ha (see Fig. 3.5 below). Only 2.7% of the land area remained as woodland.

The cleared land was mostly used for farming, but there were periods when farming was unprofitable and land was abandoned or planted with trees (Thirsk, 1957). There were also fashions for landscaping the grounds of large houses or planting coverts for foxes and pheasants, and new woodland was created on old farmland for these reasons too. In this way a scattered population of secondary woods came into existence over the last 400 years, which now covers 2644 ha (2.9% of the land area) (Peterken, 1976). These woods grew up on all kinds of farmland, mainly ploughland, but also former meadows, pasture and heathland.

The history of all woodlands within the study area was determined by reference to historical sources, old maps, and topographical relationships (Peterken and Game, 1981; 1984). Each fragment of woodland was classified as either ancient or recent secondary.

FLORISTIC SURVEY

A list of 'woodland vascular plants' was constructed for the study area, including shade casters, shade bearers and edge species, but excluding grassland and other species, even though these were common on many woodland rides (Peterken and Game, 1984). Then lists were made of woodland plants from all 89 ancient woods and from 273 secondary woods. We were unable to gain access to a few secondary woods. Although all tree and shrub species were recorded, only herbs, small shrub species and climbers are included in the following analysis.

The species lists were unlikely to be complete, but the intensity of survey was roughly the same in each wood. A test of completeness led us to conclude that the lists were 80–90% complete.

ANALYSIS

Secondary woods were divided into two groups, those that grew on land next to ancient woods (adjacent secondary woods) and those that were separated from ancient woods by a tract of non-woodland (isolated secondary woods). Secondary woods were also classified according to their period of origin. Using old Ordnance Survey maps, those originating before 1820 were separated from those originating in 1820–1887, 1887–1946 and after 1946.

The number of species in each wood

The number of species in ancient woods increased as woods got larger (Fig. 3.1), although there was much variability that could not be explained. Recent woods showed a similar species-area relationship, but above 10 ha the number of species was less than the number in ancient woods for any given size of woodland (Fig. 3.2). Below 3 ha there was little difference. Even the smallest secondary woods had an irreducible minimum of 10–20 species. Adjacent secondary woods came between the two (Fig. 3.3). The fitted regressions of the three groups of woodland were significantly different.

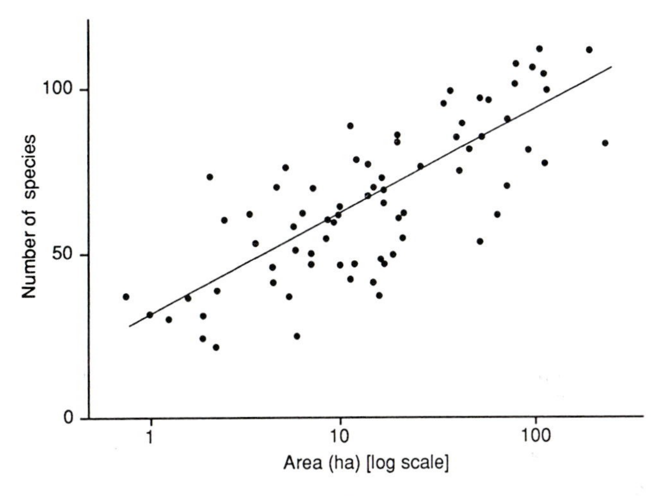

Fig. 3.1. The relationship between woodland area and number of species for the 89 ancient woods in central Lincolnshire. (From Peterken and Game, 1984.)

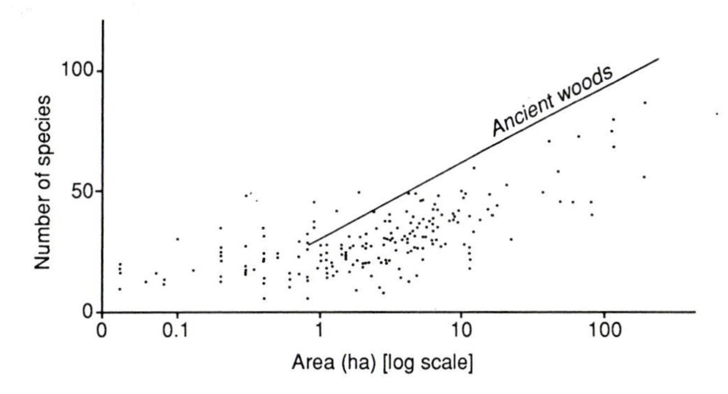

Fig. 3.2. The relationship between woodland area and number of species for 191 isolated recent woods in all regions and all periods of origin in central Lincolnshire. The species-area regression line obtained for ancient woods in central Lincolnshire (Fig. 3.1) is included for comparison. (From Peterken and Game, 1984.)

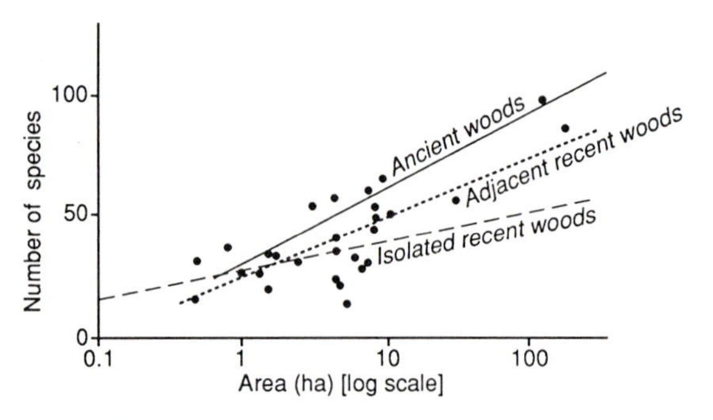

Fig. 3.3. The relationship between area and number of species for adjacent recent woods in central Lincolnshire. The species-area regression lines for ancient woods and isolated recent woods are shown for comparison. From Peterken and Game (1984).

Species in secondary woods

When examining those woods that grew up on farmland in isolation from ancient woods (i.e. excluding adjacent secondary woods and all secondary woods on former heathland), the most frequent species were those listed in Table 3.1. Without exception these are common and widespread throughout the British lowlands. They are essentially species of hedges, grassland, waste ground and gardens.

Very few species actually preferred secondary woods: most simply occurred in all types of woodland, although in ancient woods many were most abundant on the margins and disturbed ground. The four species that were positively associated with secondary woods (Table 3.2) were all widespread weeds of alkaline soils.

Do new woodlands get richer as they get older?

The species *versus* area relationships of different age classes of secondary woods (Fig. 3.4) were compared. Those originating before 1820 and those originating in 1820–1887 had an almost identical relationship. Those originating in 1887–1946 had more species in smaller areas, but the difference in slope between these and earlier woods was not significant. Woods originating after 1946 generally had fewer species, but the number of sites available was too few to test statistically.

The plant species in secondary woods appeared to build up quickly, but after about 100 years there was no further increase in total number of species. This was surprising, because some plants, such as *Mercurialis perennis*, were more likely to be found in older secondary woods than in younger secondary woods

Table 3.1. Vascular plant species found most frequently in secondary woods in central Lincolnshire. The proportion of all secondary woods in which each species was recorded is shown in 10% bands.

Band	Species
	Galium aparine
90%	
	Rubus fruticosus
	Urtica dioica
	Glechoma hederacea
	Geum urbanum
	Poa trivialis
80%	
	Heracleum sphondylium
	Moehringia trinervia
	Stachys sylvatica
70%	
	Anthriscus sylvestris
	Geranium robertianum
	Hedera helix
	Ranunculus repens
60%	
	Veronica chamaedrys
	Viola odorata
	Arctium minus
	Epilobium angustifolium
	Cirsium vulgare
50%	
	Rumex sanguineus
	Brachypodium sylvaticum
	Arrhenatherum elatius
	Stellaria media
	Dactylis glomerata
	Dryopteris dilatata
	Dryopteris filix-mas
	Rumex obtusifolius
40%	

Table 3.2. Vascular plant species that were strongly associated with secondary woodland in central Lincolnshire. These species have at least 75% of their localities in secondary woods. Secondary woods on former heathland have been excluded.

Species
Aegopodium podagraria
Lamium album
Veronica hederifolia
Viola odorata

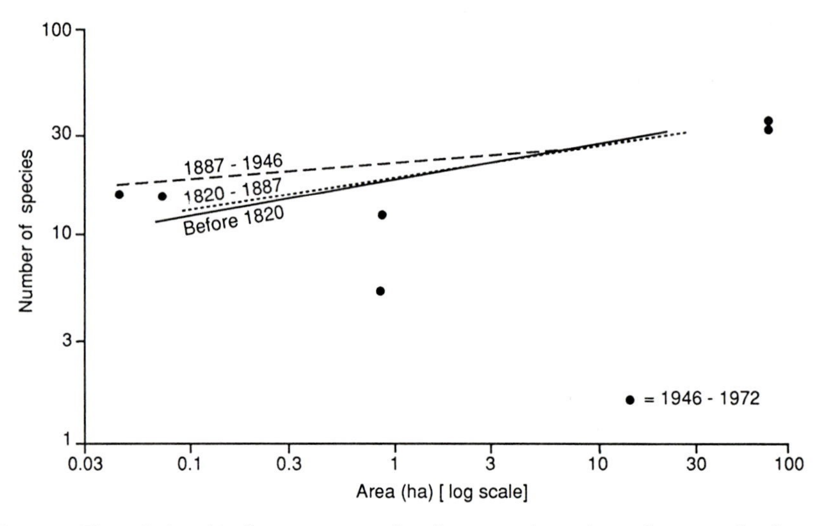

Fig. 3.4. The relationship between woodland area and number of species for four age classes of isolated secondary woodlands in central Lincolnshire. The fitted regressions for ›1820, 1820-1887 and 1887-1946 are given without data points. The six woods originating 1946-1972 are shown individually. Redrawn from Peterken and Game (1984).

(Peterken and Game, 1981). Furthermore, a study by Woodruffe-Peacock (1918) of a secondary wood near our study area showed that species continued to colonize through time by a variety of mechanisms. Presumably, some of the early colonists die out later.

Ancient woodland species

Ancient woods are richer than secondary woods because some species rarely colonize secondary woods. Ancient woods made up 25% of the sites listed and 49% of the woodland area, so if a species was randomly distributed through all woodland about 25–50% of all its locations should have been in ancient woods. In fact, 70 species (38% of all woodland species in the study area) were biased in their distribution towards ancient woods; that is, more than 50% of their localities were in ancient woods.

Table 3.3 lists the 30 species which were most strongly associated with ancient woods. Most must have been components of the original forest cover, which, for a variety of reasons, could not normally colonize new woodlands.

Ancient woodland plants in secondary woods

Nevertheless, most ancient woodland species occurred in a few ancient woods, and most secondary woods contained a few ancient woodland species. Many

Table 3.3. Vascular plant species strongly associated with ancient woodland in central Lincolnshire. Only the 30 species with 80% or more of their localities in ancient woods are listed. All species occur in at least nine sites, unless stated otherwise.

Anemone nemorosa
Calamagrostis canescens
Campanula trachelium (6 sites only)
Carex pallescens
Carex pendula

Carex remota
Carex strigosa (4 sites only)
Chrysosplenium alternifolium (4 sites only)
Convallaria majalis
Dipsacus pilosus

Equisetum sylvaticum (2 sites only)
Galium odoratum
Lamiastrum galeobdolon
Lathraea squamaria (1 site only)
Lathyrus montanus (3 sites only)

Luzula pilosa
Luzula sylvatica
Lysimachia nemorum
Lysimachia vulgaris
Maianthemum bifolium (1 site only)

Melampyrum pratense
Melica uniflora
Milium effusum
Neottia nidus-avis
Oxalis acetosella

Paris quadrifolia
Platanthera chlorantha
Potentilla sterilis
Vicia sepium
Vicia sylvatica (3 sites only)

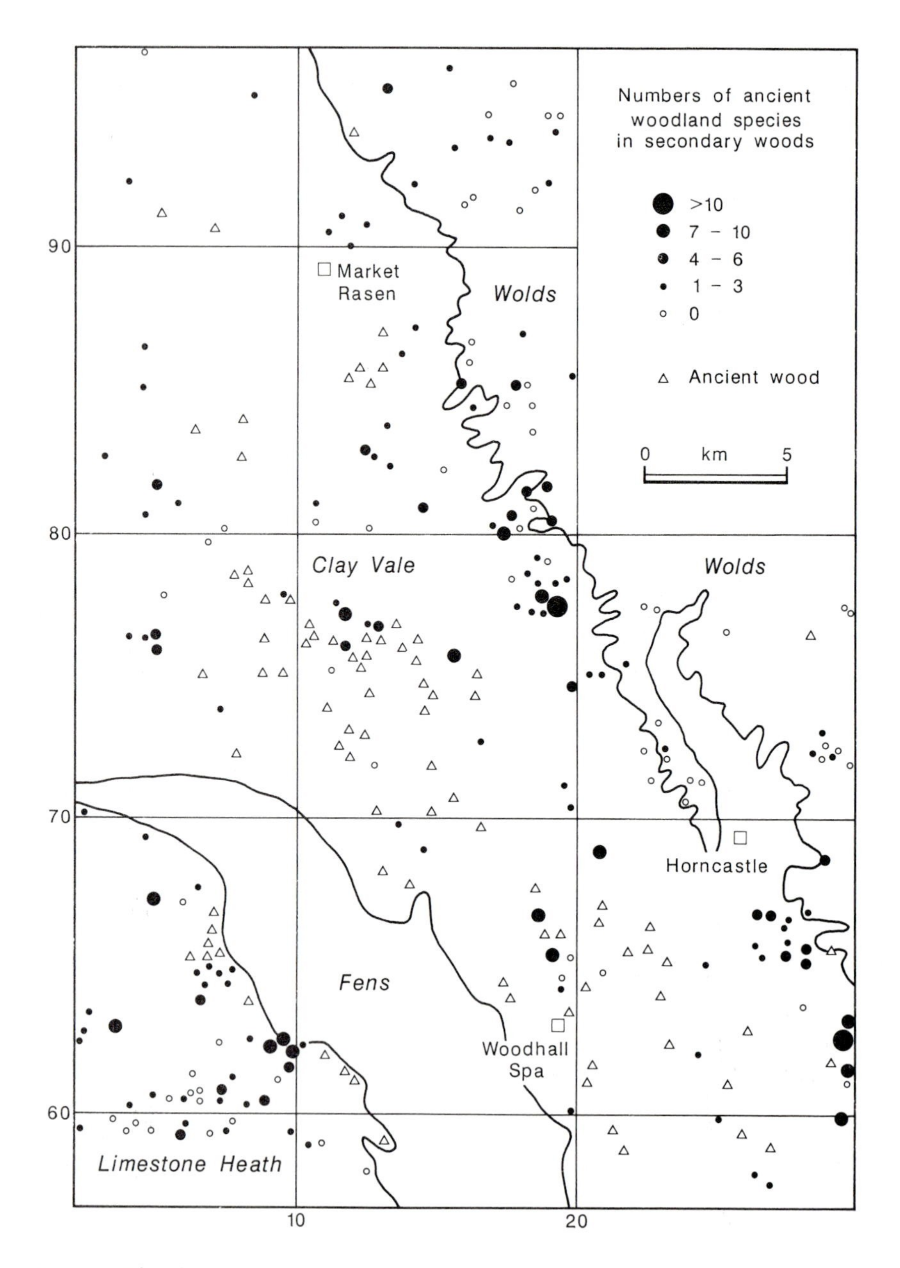

Fig. 3.5. The distribution of ancient woodland species in central Lincolnshire woods. The number of ancient woodland species in each wood is indicated.

were in adjacent secondary woods, that is, they had just 'hopped over the fence', but some were in isolated secondary woods. How did these species get into these woods?

Figure 3.5 shows the location of isolated secondary woods and how many of the 70 ancient woodland species each contained. Some had none and most had very few. Although the smallest numbers were found on the Wolds and Limestone Heath (well away from the concentrations of ancient woods) there was no obvious relationship with distance from ancient woods.

Throughout the district, a scatter of woods stood out as richer than their neighbours. The secondary woods that were richest in ancient woodland species are listed in Table 3.4. These woods tended to have one or more of the following features:

1. They were close to an ancient wood or directly linked in time through a wood-relic hedge (Pollard, 1973), or they had been close when they originated. Woodlands plants could have colonized new woodland from these ancient habitats without having to cross farmland.
2. They contained a headwater stream or were close to one. Small streams and the associated marshland commonly support woodland species, often in the semi-shaded, moist conditions also found in woodlands.
3. They grew on land beside or astride a parish boundary. These ancient boundaries often had substantial banks and hedges that provided stable refuges for woodland plants. Unfortunately, so many had been destroyed between 1945 and the start of the study in 1970 that their value as woodland habitats could not be properly assessed.
4. They covered a relatively large area. (The median secondary wood was about 2 ha.) Larger woods tend to cover a wider range of former land uses and touch a larger number of hedges, so their chance of being colonized by a particular woodland species is greater than that of smaller woods.
5. They originated before 1887. This means that woodland plants had had more than 100 years in which to colonize.

Every site in this list had at least one of these features and most had three of four. The poorer woods generally had one or none of these features.

The occurrences of ancient woodland species in isolated secondary woods were thus exceptions that proved the rule. The ancient woodland species were present because these woods had not in fact been ecologically isolated. They had perhaps long survived in a wood-relic hedge, colonized only a very short distance across unwooded territory, or survived in marshland, streamsides or ditches. This inference is supported by the presence today of such species in the few remaining examples of these habitats, even after 40 years of intensive modern farming. It is also possible that some of the secondary woods with many ancient woodland species were growing on land that had been pasture or meadow, rather than arable, before it reverted to woodland, but this has not been properly investigated.

Table 3.4. Characteristics of secondary woodlands in central Lincolnshire containing six or more ancient woodland species.

Secondary woodland	Area (ha)	Period of origin[b]	No. of ancient woodland spp.	Relation to ancient wood[a]	Stream through/ alongside wood	Parish boundary through/ alongside wood
Clay Close Holt	53	4	15	—	A	P
Wilksby Plantation	12	3	13	*	—	P
Stixwould Plantation	38	2	10	w	A	P
Sotby: Copt Hills and Brant Hills	22	3	10	—	A	P
Grantham's Cottage	6	4	10	*	—	P
Top Plantation	15	4	9	—	—	—
Metheringham Barff Farm East	>1	1	7	W	A	—
Metheringham Barff Farm West	>1	1	7	W	A	—
Metheringham Barff	3	2	7	W	A	—
Curtois Holt	7	4	7	*	—	P
Edlington Moor Farm	4	3	7	w	—	P
Goltho Pond	1	2	7	w	A	—
Coultas Wood	7	3	7	W	A	—
Panton Park	9	4	7	—	A	—

Revesby West Lane	18	3	7	—	A	P
Edlington Scrubs	20	4	7	—	—	P
Revesby Long Strip	8	3	7	*	—	—
Blankney Brickpits	5	2	6	—	A	—
Sudbrooke Park	4	4	6	—	A	—
Snarford Holt	3	4	6	—	A	—
Black Plantation	5	2	6	w	A	P
Lodge Covert	21	4	6	—	A	P
Fishpond Wood	1	4	6	—	A	—
Collow Holt	3	4	6	—	—	P
Croppersgorse Plantation	16	4	6	—	—	P
Normanby Dales	14	3	6	—	A	P
Guide Post Plantation	<1	3	6	w	—	—

[a]W = direct link with ancient wood (e.g. wood-relic hedge); w = <500 m from an ancient wood; * = 500–600 m from a large ancient wood.
[b]4 = before 1820; 3 = 1820–1887; 2 = 1887–1946; 1 = after 1946.

Summary of Findings in Central Lincolnshire

Although much of the evidence is circumstantial, the broad picture that emerges from 300 years or more of 'set-aside' in central Lincolnshire can be summarized thus:

1. New woodlands on old fields quickly develop a flora of mainly hedgerow and garden 'weeds'.
2. They do not develop a rich woodland flora containing all the appropriate species of the original woodland, and will not do so even if they are given another 300 years.
3. This is because almost two-fifths of the original woodland flora is unable to colonize new woods, or does so only slowly and across short distances. There must have been few occasions when ancient woodland species have colonized across a large stretch of hostile territory.
4. New woods on old fields can develop a relatively rich flora if they incorporate habitats with some direct link in time and space with the original woodlands such as wood-relic hedges, naturally meandering streams, wet meadows and pastures.

Similar Observations from Other Regions

Ancient woodland species of many kinds have been identified in many parts of Britain (Peterken, 1981). They include vascular plants, bryophytes (Ratcliffe, 1968), lichens, insects (Harding and Rose, 1986; Chapter 2) and molluscs (Boycott, 1934). They have also been identified in the USA (Whitney and Foster, 1988), The Netherlands (Hermy and Stieperaere, 1981) and Poland (Dzwonko and Loster, 1989). In many parts of Britain the woods growing on what was farmland can be rapidly distinguished by their flora from the ancient woods.

On the other hand, as one might expect, the difference between ancient woods and recent woods is much less where woodland density is high and farming has been mainly pastoral, with little cultivation. This supports the general diagnosis that this is a phenomenon of ecological isolation and soil changes brought about by cultivation.

Future Development of the Flora of New Woodlands

Any new woodland that is planted, or allowed to grow naturally, on what is now farmland will almost certainly replace arable crops or ley grassland, simply because there is very little other land left and conservationists want the few remaining patches of heathland, old grassland and wetland to remain unplanted. Most of this land will have been ploughed, fertilized and drained in recent years.

Furthermore, in most parts of lowland Britain a high proportion of the hedges have been removed, most streams have been straightened into ditches and even roadside verges have become so eutrophicated by fertilizer drift from nearby fields that they support only a limited range of rank grasses and tall herbs (Peterken and Hughes, 1990). Moreover, during the last 50 years, 8% of all ancient woodland – the principal refuges of the native woodland flora and fauna – have been destroyed by clearance (Spencer and Kirby, 1992) and many others may have been impoverished by coniferous plantations.

The new woodlands will therefore be established in a floristically degraded environment. The plants that will colonize first will usually be common and widespread species that grow strongly on dry, alkaline, enriched soils and woodland species with efficient dispersal mechanisms. Other woodland species (that is those of wet, acid or non-enriched soils, many of which colonize slowly) will be unable to colonize because there are now few relict locations from which they might spread and any propagule that does reach a new woodland will face overwhelming competition from vigorous herbs already established. In central Lincolnshire, in the future, even more than the 40% of the woodland flora which has not been able to colonize new woodland effectively for the last 400 years, may be restricted. However much one may wish otherwise, few of the new woods will acquire a full representation of woodland species, and the species that will not be there include some of the most characteristic and attractive.

If we wish these new woods to develop fully representative floras, we must site them close to ancient woods, crop the land for a few years without fertilizing, and plant or seed in those species that cannot readily colonize (Francis *et al* 1992). Where we lack ancient woods or ancient semi-woodland habitats and the soils cannot be stripped of nutrients, we must plant not only the trees but the whole woodland ecosystem.

Acknowledgements

Much of the research on which this chapter is based was undertaken jointly with Margaret Game. Roger Bolt helped to prepare Table 3.4 and Figure 3.3.

4

Depopulation and Afforestation: Sources and Methods from 19th Century Italy

F. SALBITANO

SUMMARY

The natural regeneration of woodland in abandoned fields is a quite widespread phenomenon in the mountain regions of developed states of the West. This chapter shows the relation between depopulation, social changes and ecological variations as historical process in two case studies in north-east Italy. By focusing on 19th century cadastres, the study tries to analyse the use of historical records as key sources to understand the ecological history of the area at different levels. Afforestation is strongly linked to depopulation and the way secondary succession develops is partially dependent on the previous land use. The information derived from the General Cadastre, combined with other written sources, contributes to the interpretation of the process of secondary succession at the landscape level. By analysing the more specific records it is also possible to provide information at a site level, although these data must be cross-referenced with other historical, archaeological and ecological information.

INTRODUCTION

The natural regeneration of woodland in old, abandoned farmland is the consequence of diminishing human control over agricultural ecosystems. These secondary successions have been one of the most studied issues in the field of terrestrial ecology over the last 20 years or so in both Europe and other western countries. The ecological effects of secondary succession include substantial changes in plant population structure both during and after the abandonment. Several studies have shown the importance of a knowledge of site history, including previous forms of land-use and land-management practices in order to understand the dynamics of the recolonization processes. The old agricultural and pastoral landscapes were very rich in elements, such as isolated trees, small groups of trees and shrubs, planted trees for cultivation and the provision of fodder, walls, hedges, and so forth. These elements of the landscape frequently provided suitable sites for future plant generations (Harper, 1977).

The aim of this chapter is to analyse a series of specific historical records, the 19th century cadastres, and to examine their potential as a possible information source for ecological history case studies. In particular, the information given by the cadastres is used to contribute to our understanding of human influences on woodland management and the relationship between politics, socio-economic events and ecological change in two prealpine regions in Eastern Italy: Frisanco, in the Carniche Prealps (Province of Pordenone), and Taipana in the Giulie Prealps (Province of Udine).

This chapter concentrates on the following:

1. The main social, economic and political changes that influenced migration and depopulation.
2. The relationship between depopulation and changes in woodland area and woodland type.
3. The 19th century landscape, land-use descriptions and human activities before the period of high population in the late 19th century.
4. The difference between the descriptions of land-management practices given in official documents for the two areas and the Cadastre classification system. This allows a critical assessment of the value of the sources to be made.

SOURCES AND METHODS

Nineteenth century cadastres have frequently been used as a simple and readily available source, sometimes almost like a magic wand, to reconstruct past landscapes and understand social and economic problems. Frequently the final published records, the tax and census registers, are taken as a picture of reality (Corbellini, 1986). However, when the complete set of documents, from the preliminary surveys through to the final records, are considered, it is clear that the cadastres are a very rich yet sometimes contradictory source of information.

In the case of the Italian cadastre census, this complexity is linked to the history of the modern geometric cadastres. The first methodical attempt to prepare geometric descriptive cadastres for several Italian regions was made at the beginning of the 19th century during the period of Napoleonic rule. The name given to the project by the French Administration was the General Kingdom Cadastre, referring to the political status of Italy within the Napoleonic Empire as the Kingdom of Italy. A large amount of classification and mapping was carried out during the brief period (1805–1813 in these regions) of French power, but the preparations for the cadastre were left incomplete. The restoration governments sometimes completed the cadastre by building on the work carried out for the French cadastre; in other instances they used similar methods to develop new cadastres. The latter was the case for the Austrian Empire administration in the Lombardo-Veneto Kingdom of north-eastern Italy, which Taipana and Frisanco municipalities belonged to until the unification of Italy (in 1860–1866). The links between the French cadastres and the later documents are emphasized by some confusion in the records. For example, all the maps officially used after 1830 are entitled *Catasto napoleonico* and the Archive Records are called 'Napoleonic and Austrian Cadastres'.

The cadastral survey of north-eastern Italy was started in 1807 for the Napoleonic Government of the Passariano Department. The work began by mapping and by establishing the borders. When French power ended in 1813 only a few of the maps had been completed. However, Corbellini (1986) points out that there was a substantial level of agreement between the French and Austrian administrations concerning the way in which 'a modern state of European background' should be managed. All the preliminary documents follow the same approach as the Napoleonic cadastre and are preserved in the National Archive of Venice (ASV).

A comparison of the preliminary documents, finished between 1826 and 1827, and the final cadastre records (National Archive of Udine, ASU), active from the 1851, shows the following differences. The Austrian Empire surveyors defined the land-use categories that were used in the preliminary survey according to the previous French survey. The number of categories is very high: there were 78 in Frisanco and 172 in Taipana; the description of the territory is precise and methodical. The 1851 Records of the Austrian cadastre, that is the official census documents until 1920, make use of considerably fewer land-use types; there were 56 categories for the entire Friuli region. This was much more workable from the point of view of the taxation system operated by the Austrian Administration and later by the Kingdom of Italy. The heavy burden of taxes was one of the main causes of the economic depression for these and other isolated mountain regions.

The preliminary documents of the Austrian cadastre include different sections: General Description of the Territory (DGT, ASV); Detailed Agricultural Information (DAI, ASV); Table of Tax classification per land-use type (TT, ASV); and Summary Table of land-use area per type and number of land units (ST, ASV). The 'Sommarioni' cadastre records (SOM, ASU), and the General Cadastre (CG, ASU) in ink manuscript, show the census plot list, land-use type, number of the land owner and plot area. Generally, pencil amendments show changes made up until 1925, but the changes in the two case-study regions have only been noted up until 1909. The landownership book (OB, ASU) lists the name, place of origin and ownership types (Public, Private, Church), with the corresponding code in the cadastre records. The latter document, together with the census of population, general enquiry and statistics, were used to study population dynamics and social changes during 19th and early 20th centuries.

DISCUSSION

Social and economic changes in the 19th and 20th centuries

The Slavic people of Taipana had had very little contact with either the central administrations or neighbouring regions since the Early Middle Ages when the Slavs colonized a large mountain region in eastern Italy. The Taipana population had a long-standing argument with the Venetian Administration during the 18th century in which they defended the traditional common use of the public pastures. This common use remained unchanged until the beginning of the 20th century (Mirmina, 1985). The first important trading between Taipana and Udine took place during World War I, when the firewood supply crisis forced the Municipality of Udine and the Italian Army to buy wood from Taipana. The population of Taipana reached its peak around 1910 and it became impossible to meet the demand for wood because of the scarcity of wood to cut.

The surveyors in 1826 describe the Frisanco and Taipana people as living in very poor conditions, with low crop yields, very bad health conditions and a total lack of tracks or roads – even mule tracks – to carry the few agricultural products to open markets. People walked the distance of roughly 30 km to Udine to sell hay, which they carried on their shoulders. Similar conditions prevailed from after the Unification of Italy up until as late as the 1930s.

Land use and agricultural products

According to the preliminary descriptions in the cadastre, the main agricultural products of Taipana were wheat, sweetcorn, potatoes, hay, cheese and firewood. The quality of cheese and wheat was very low, sweetcorn, hay and firewood were of medium quality and potatoes were of high quality. Only the hay was marketable; the rest was consumed at home. The farmers did not use the plough, but only the spade. The land-use type *aratorio* (tilled by plough), was not used in either the Taipana or the Frisanco cadastres. In the preliminary and final records, the surveyors used the term *coltivo da vanga* (cultivated by spade). No rotation or fallow-field methods were practised.

The only unusual improvement noted by the Austrian surveyors was the plantation of softwood trees in cultivated fields or meadows to fertilize the ground. This is a special usage of the term softwood: the trees described here were willow and alder and not coniferous species. The practice was described only in the preliminary documents for Platischis in the eastern part of the region. The 1826 descriptions of the Frisanco area are quite similar to the earlier documents apart from the extent of bean production. The number of cattle and amount of wild grazing in the public pastures was low. Pollarding and pruning of oak, ash and maple to produce fodder were also described.

During the 19th century several relatively minor land-use changes took place. In Platischis (Taipana), for example, the number of sheep-breeding farms decreased and the number of sheep declined from 2000 to 276. On the other hand, the number of cattle increased and they became an important source of income towards the end of the 19th century. Up until 1909 the number of new *stalle con fenile* (cowsheds and haylofts) and of **prato** (meadows) mentioned in the cadastre revisions increased. The corollary was a decrease in the number of times **zerbo** (uncultivated shrub land) and **boschina** (4–6 years short-rotation coppices) were mentioned. The land-use types that showed the greatest increase in the 19th century were *sterile* (arid), and *sasso nudo, ghiaia nuda, rupe nuda* (bare stone, rock and gravel). At the beginning of the 20th century, therefore, when the highest resident population was registered both in Frisanco and in Taipana, the land-use that became increasingly mentioned was arid and bare ground.

Population variations

During the greater part of the 19th century there was little migration from the whole of the prealpine region. From 1826 to 1834 there are no records of population movement either from and or to Taipana and Frisanco. Around the 1840s and 1850s emigration was only occasional and the proportion of migrants to the resident population as a whole was less than 1%. Population growth remained high until the end of the century. The highest populations were reached during the decade 1871–1881 in Frisanco and during the decade 1901–1911 in Taipana.

Towards the end of the 19th century and at the beginning of the 20th century emigration on a massive scale began. This was caused by a wide range of factors including:

- a high rate of population increase at a time when technical changes were relatively stagnant;
- a decline in soil fertility and an increase in soil erosion;
- a fall in the value of agricultural products;
- heavy taxation by the Central Administration – especially a fiscal policy that penalized the low incomes; and
- a lack of private and public investment.

Changes in property ownership also encouraged depopulation. Over the 19th and 20th centuries, the legal changes that brought about a decline in common land and Church property had the additional effect of increasing the fragmentation of land holdings. Plot sizes became smaller and the holdings of individuals became more dispersed. This was a very significant change for the Slavic area of the prealps where a long-standing tradition of common grazing rights had developed over the centuries. In 1826, the average area per cadastre plot was 0.6 ha in Taipana (0.3 ha in 1990), while in Frisanco it was 0.3 ha (0.18 ha in 1990).

People frequently migrated within the region but emigration was more common. The migration started earlier in Frisanco than Taipana where the birth rate dropped during World War I and migration increased in the 1920s (Valussi, 1971; Prost, 1977). At the end of the 1980s the average age of the resident population was twice that of the 1940s. The depopulation resulted in much abandonment of land and a heavy reduction in farming. The number of grazing sheep in the village of Platischis (Taipana Municipality), for example, declined from 3000 early in the century to none in 1980. The lack of grazing has meant that woodland has recolonized the greater part of the prealps within half a century.

Descriptions of the forest landscape and change in the 19th century

Today woodland is the predominant land-use in the prealps. Observation of changes in the woodland area over time using sources such as 19th century cadastres, the forest map of Italy and technical maps of mountain communities and regional administrations show the strong relationship between depopulation and the increase in woodland (Fig. 4.1). This is hardly surprising; I am more interested, however, in the following questions. What woodland types are involved in the process of afforestation? What changes in woodland type have taken place during the period? What differences are there between the classifications used by various surveyors and the real land-use types and land-management practices, and finally, what is the significance of these differences in terms of our interpretation of landscape and ecosystem changes?

I will now try to answer the historical questions, focusing on a critique of the sources and whether they are relevant or not to an interpretation of secondary succession dynamics and the ecological effects of old-field colonization. The cadastres classify trees into one of four main types: hardwood broadleaves (e.g. oak, beech, elm, maple, cherry, ash and hornbeam); softwood broadleaves (e.g. alder, willow and lime); hardwood conifers (e.g. larch, pine and juniper) and softwood conifers (e.g. Norway spruce and silver fir).

I have already pointed out that both the Napoleonic land census and the preliminary documents of the Austrian cadastre submitted to the Tax Commission in 1826, describe land-use in a very detailed manner, and that land-use is simplified in the later cadastre documents. For example, the precise category *pascolo con piante dolci* (ST, ASV) (pastures with softwood trees, e.g. alder, willow) from the earlier documents is subsumed into the more general category *pascolo boscato ceduo misto* (CG, ASU – pasture and mixed broadleaved coppice) in the later versions. This general category will also include areas described in the earlier version as *pascolo con piante forti* (ST, ASV – pastures with hardwood trees (e.g. beech, elm, maple, ash and hop-hornbeam) and *pascolo con piante cedue forti* (ST, ASV – pastures with coppice hardwood stools).

In general, the qualitative comments and descriptions included in the preliminary documents were correctly reclassified into the general categories used in the final documents. Sometimes, however, there are descriptions that the

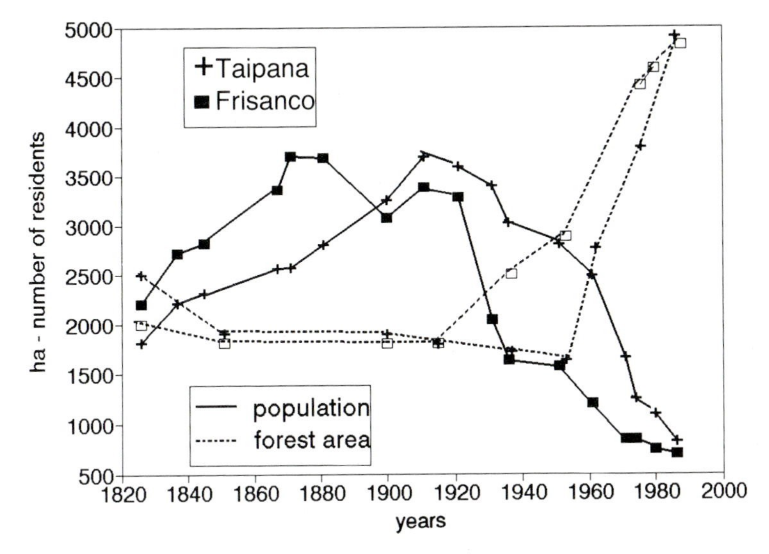

Fig. 4.1. Changes in population and forest area in Taipana and Frisanco 1820-1990.

surveyors cannot place within the criteria given by the central authorities. Some of the qualitative comments are adversely critical of the condition of the woodlands: in one instance the surveyors noted the 'reprehensible status and utilization of the prealpine woodlands' (DGT, ASV). The surveyors also provide detailed notes on the cultivation systems used, production and the quality of regeneration.

According to the general classification of 1851 (CG, ASU), the principal woodland types were:

1. **Coppice.** The stools were generally cut at the base but pollarded trees were included in this category. Three subtypes are given according to the species composition: (i) hardwood broadleaves: oak, beech, elm, maple, cherry and ash; (ii) softwood broadleaves: alder and willow; and (iii) a mixture of hardwood and softwood species.
2. **Chestnut coppice.**
3. **Short-rotation coppice.** Tree and shrub species coppiced on a 3–4 years rotation.
4. **High forest.** The woodland area was utilized for leaf (fodder) and timber production. Four types based on the tree species composition are described: (i) beech high forest, which had a longer rotation than the hardwood coppices – only the biggest trees were cut while the smaller ones were left for the 'forest regeneration' (DAI, ASV); (ii) hardwood species such as hardwood coppice; (iii) softwood species such as poplar, alder, willow and lime; and (iv) stands of mixed hardwood and softwood.

5. Conifer stands. Two types are described: (i) hardwood: larch, pine, juniper, and so on; (ii) softwood: Norway spruce, silver fir; and (iii) mixed.
6. Chestnut woodlands. Chestnuts cultivated for nut production.

Three additional land-use types were strongly connected with trees and woodlands. The categories depend on the density and structure of the tree or shrub layer. The descriptions given by the preliminary documents help to clarify these categories:

1. Sparsely wooded pasture. Pasture with occasional coppice stools and a discontinuous tree canopy. Three different types are given, according to the species composition as in the categories for coppice.
2. Bushy pastures. Pastures with scattered bushes and scrub, including species such as juniper, brooms and briars.
3. 'Zerbo'. Unfruitful weed- and bush-covered land.

By combining these general categories of 1851 with the data from the preliminary documents (DAI-TT-ST, ASV) it is possible to distinguish the following principal woodland types.

1. Coppice. Most of the coppice woodland consisted of mixed hardwood and softwood broadleaved species. Also found within this type were beech coppices, sparsely wooded pasture and the wooded banks of rivers and streams. The coppices grew on very steep slopes about 5 km from roads and water courses. The cut poles were carried by shoulder from the woodland to the villages (DAI, ASV). There was no regular cutting system, but each year the people cut the oldest and biggest stools. Every 15 years the remaining stools were also coppiced, leaving few plants uncoppiced (TT, ASV). Very few of the stools died; there was frequent natural regeneration and no need to plant trees or make special efforts to retain standard seed trees (DAI, ASV). The density of stools was low. It was pointed out that 'the wood can survive if care has been taken during the cutting and is preserved by animal browsing' (DAI, ASV). There were three census classes related to stand density, soil type, the distance from the village and productivity. Coppices of the first-class quality were found close to the village and had no soil erosion (TT, ASV). This class had a productivity of 53 m^3 after 15 years' growth. The second and third classes had a lower density of stools than the first class and were found at a greater distance from the village. The second class had similar soil to the first class and had a productivity of 40 m^3 after 15 years. The third class had less-fertile soils and suffered from erosion and landslides. It had a productivity of 27 m^3 after 15 years' growth (TT, ASV).

2. Beech high forest (only in Taipana). An auction sale of the municipality beechwoods of Taipana took place when the mature stands were roughly 100 years old. The cut was selective and the trees that were to be felled were marked by the Forestry Commission. Smaller trees up to 20 years of age were left uncut. The yield was about 150–200 m^3/ha of wood for charcoal production and firewood.

3. Conifer forest (only in Frisanco). Five plots of silver fir forest were present in Frisanco territory. They were in a very bad structural condition with a low density of trees. Many trees were diseased. No particular form of management was practised. The people cut the biggest trees and collected wood from dead trees.

4. Short-rotation coppices of mixed broadleaves. Although these coppices consisted of both softwood and hardwood broadleaves, softwoods predominated. These coppices were scattered over the communal area but especially along the river banks. The rotation was 6 years (DAI, ASV). The stools lived for 35–40 years and after cutting resprouted easily. They needed to be protected from browsing animals, especially goats (DAI, ASV). The coppices were classified on the basis of stool density, species composition, location and soil nature. The first and second classes had a high number of hazel, alder and ash stools, with second-class stands being more distant from the village. Third-class stands had a lower density of stools, more sandy soils and suffered from landslides (TT, ASV).

5. Meadows with scattered broadleaved coppice stools and trees. Scattered hardwood and softwood broadleaved stools in the meadows were coppiced every 15 years. The wood production was estimated at roughly a quarter of that of the coppices (13 m^3) (TT, ASV). This type also included bushy meadows, which were used to grow faggots and cut every 3 years.

6. Pastures with scattered broadleaved coppice stools and trees. Low-density stools in the pastures were cut every 8 years. The products were firewood or stakes for bean cultivation (DAI, ASV).

7. Pastures and wild chestnuts. Chestnut groves in pastures had a density of 35–40 trees/ha. Much wind damage was observed. The chestnuts were grown for poles and timber and not for nuts. It was impossible to market the poles and timber because of the poor roads (DAI, ASV).

Having considered the characteristics of the principal types of woodland it is now possible to make some broader generalizations about the woodlands of the area based on the cadastre evidence. Most of the woodland consisted of hardwood broadleaved coppice and beech was the most frequent species. There was considerable spatial variation in the density of stools, the size of individual woods and the distribution of species.

The selection cuts created uneven-aged and irregular stands in both coppice and high forest. It is rather difficult to understand the regeneration regime. Regeneration was always mentioned as a very easy, natural process. According to the observations of the surveyors there was no need to plant trees or to keep standard trees to produce seeds. So far as stool replacement was concerned, it was simply recommended that care should be taken of the small shoots or saplings left after the cutting. It was never stated whether seed regeneration took place. In the descriptions of high forest the individuals left after the marking were often vegetative shoots rather than maiden trees. The 'high-forest' classification appears to have been based on the structure of the woodland and the timber produced, rather than on the regeneration system used.

Even if the surveyors recommended regular cutting and tried to regulate the land-use situation, in reality the land-management practices were quite confused. The woodland production was obviously heterogeneous in order to support the different requirements such as firewood, faggots, branches, leaves, stakes, posts, timber and charcoal. In consequence, the distinction between wooded and unwooded landscapes was always rather vague. It is very difficult to know whether some types, such as the meadows and pastures with scattered broad-leaves, should be classed as woodland or agricultural land.

The landscape of the two study areas is shown by the cadastre to have been extremely rich and various in different elements. By cross-referencing the broad 1851 classes with the more specific classes derived from the French categories, it is possible to provide an indication of the extent of the complexity for Taipana (Table 4.1). The list of elements given in the table could be extended to include additional categories such as nuts, hazel shrubs, and meadows with hedges.

Evaluation of forest landscape change: the example of Taipana

During the last two centuries the woodland area of Taipana increased from 1900 ha to 5000 ha. The proportion of woodland rose from 29 to 80% (Salbitano,

Table 4.1. The land-use mosaic in Taipana.

Detailed categories in ST, ASV	Number of landlots	Total area (ha)
A: Land-use group in CG, ASU – meadows		
Meadows with mixed broadleaved trees	13	0.43
Meadows with hardwood high-standing trees	4	0.7
Meadows with softwood coppiced plants	3	1.0
Meadows with softwood trees	2	0.4
Meadows with fruit-bearing chestnuts	7	1.4
B: Land-use group in CG, ASU – pastures with scattered broadleaves, coppice stools and trees		
Shrubby pastures with hardwood trees	1	0.2
Pastures with hardwood trees	1	0.2
Pasture with softwood broadleaved trees	3	0.2
Unproductive shrubland with softwood broadleaved trees	3	0.3
Unproductive shrubland with hardwood broadleaved trees	2	0.01
Pasture with timber chestnuts and so on	1	0.1

1987). The area of beech woodland has remained remarkably constant over the period. A policy of artificial coniferous afforestation was introduced in the 1930s, and today such conifer plantation covers 12% of the total land surface of Taipana. Most of these plantations were established as protection forest on areas of land classified in the cadastre revisions of the early part of the 20th century as rocks and gravel. Most of the natural regeneration of woodland, in contrast, has taken place on areas classified in the cadastre as meadows, pastures, meadows sparsely wooded, pastures sparsely wooded and cultivated fields with trees. The principal species that have colonized these areas are ash, maple, black alder, hornbeam, hop-hornbeam and manna ash. These species are mixed in different proportion depending to some extent on the soil type.

Conclusions

The information provided by the cadastre can be a good historical data base that can be used to understand landscape and ecological variations at a range of levels. However, it is necessary to compare different records for the same area and to assess critically the relationship between the descriptions made by the surveyors and the actual land-use. The level of information available on particular practices, such as pollarding and pruning, and elements, such as walls and hedges, is rather low. On the other hand, the cadastre is a good source of information on production, stand characteristics, regeneration, woodland density and grazing. By comparing population data with the land-use change data provided by cadastre revisions it is possible to investigate the stage when the population of the district became unsustainable. After several centuries when the land-use system had been relatively stable, there was a sudden shift in land-use. It is clear that there was a complex relationship between central government policies, impoverished mountain areas, depopulation and drastic landscape change.

Abbreviations

ASU: Udine, National Archive
ASV: Venice, National Archive
CG: General Cadastre Documents
DAI: Detailed Agricultural Information
DGT: General Description of the Territory
OB: Owners Book
SOM: Summarization of Cadastre Records
ST: Summary Table of land-use type per land parcel
TT: Table of Tax categories

5

The Influence of Old Rural Land-management Practices on the Natural Regeneration of Woodland on Abandoned Farmland in the Prealps of Friuli, Italy

M. GUIDI AND P. PIUSSI

SUMMARY

The eastern Prealps have been intensively exploited by agriculture in the past. Since the 1950s, fields, meadows and pastures have been abandoned and several woody species have invaded them. In a relatively short time, a dense woodland cover has established itself. Most of the young stands have been temporarily classified as ash-maple and manna ash-hornbeam woodland. Within each type there is much variation. Our research shows that early colonization as well as stand structure and composition are partially related to past land-use practices. These include soil-cultivation techniques, forage production from fodder trees and meadow improvement. The methods employed in the research included description of the structure and composition of the woods using systematically distributed sample plots, oral history and the field investigation of terraces, hedges and walls.

INTRODUCTION

The depopulation of rural areas and the abandonment of land cultivation and stock raising are widespread phenomena over most of the southern Alpine slopes. These processes started some hundred years ago, dramatically increased after World War II and are still underway (see Chapter 4). Abandonment is usually followed by the invasion of woodland species and the natural regeneration of woodland.

Our research describes secondary woodlands created by this succession process on the prealpine areas of the eastern Italian Alps, in the Friuli region, and tries to understand the factors that influence it. Preliminary inquiries in two areas of this territory – Taipana in the east and Frisanco in the west – show that tree species usually appear immediately after cessation of cultivation or mowing activities, their growth is fast, and dense stands are formed in a few decades. Many species are represented in secondary woodlands and it is possible to identify different forest types.

Woodland types seem to be partially controlled by site characteristics, but during our field work we came to the conclusion that past land-exploitation practices could also have been responsible for today's situation. In fact, the use of trees, especially for animal husbandry, could have influenced the succession. It is therefore useful to know which woodland species have been managed, what the management practices were, and how these species participate in the formation of secondary woodlands.

The object of this chapter is to describe only those practices that seem relevant during secondary succession, leaving out standard horticultural practices regarding chestnut (*Castanea sativa*) and other fruit trees. Those practices that, during this first part of our research, seem to have had a relevant impact on vegetation dynamics will be described; namely alder (*Alnus glutinosa*) cultivation, leaf collection for fodder and soil cultivation. The results presented are preliminary and the research, which is still underway, has now been extended to cover a wider region.

STUDY AREA AND RESEARCH METHODS

The research covers the territory between the Isonzo and the Cellina valleys. It is bounded on one side by the plain (approx. 200 m above sea level) and on the other side by the beech (*Fagus sylvatica*) belt, situated between 700 and 1000 m above sea level. Exposure, slope and general morphology are not uniform. The climate is characterized by heavy precipitation (1500–3500 mm) with two maxima in spring and autumn; temperatures are mild.

There are two main geological formations: limestones and dolomitic limestone (from the upper Triassic, Jurassic and Cretaceous period), and marl and sandstones (from the Maastrichtian-Paleocene-Eocene period). Due to the complex tectonic situation, the two formations are mixed and overlap each other. Standard geological maps (scale 1:100 000) are therefore useless in trying to detect a correlation between the type of vegetation and the geological parent material. Unfortunately, however, detailed geological maps are seldom available and field reconnaissance is not always easy.

Research carried out in the pilot areas of the communities of Taipana and Frisanco showed that the composition and structure of secondary woodlands was influenced by old farming practices. It was therefore decided to obtain a more detailed picture through a series of interviews with farmers over the entire research area. Further field study was undertaken to discover whether old practices were influencing stand structure and dynamics, and, if so, in which way.

Old fields, meadows and chestnut groves are now occupied by different woodland types at various levels of succession, depending on how long they have been abandoned. The most important woodland types are:

- *Fraxinus ornus-Ostrya carpinifolia* woodland;
- *Fraxinus excelsior-Acer pseudoplatanus* woodland;
- *Robinia pseudoacacia* woodland.

Many tree and shrub species are present but the dominant ones are ash (*Fraxinus excelsior*) and maple (*Acer pseudoplatanus*) (Figs 5.1 and 5.2). The best stands, at 30–40 years, have a density of 21–35 m^2/ha and the height of the trees reaches 25 m and more. Seedlings and saplings under tree cover are mainly represented by *Fraxinus excelsior* and *Acer pseudoplatanus*. The rural landscape until the middle of the century was characterized by fields, meadows and

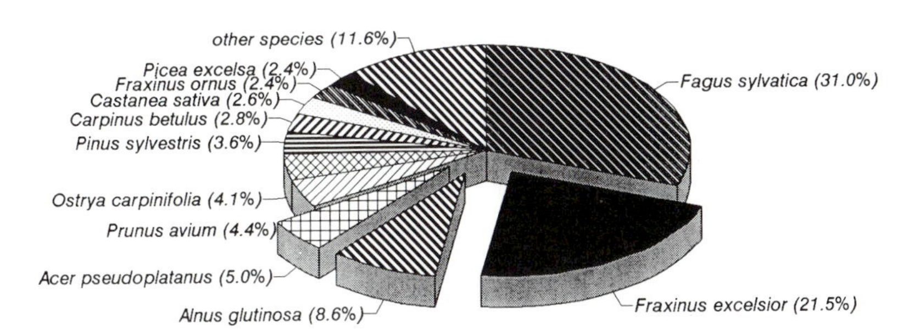

Fig. 5.1. Basal area distribution of the tree species in the Taipana territory in the eastern part of the study area. Alder cultivation was practised in this area; data are included for old fields and woodland.

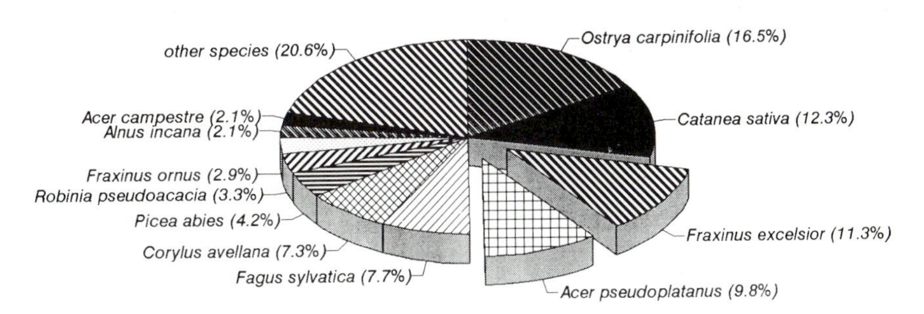

Fig. 5.2. Basal area distribution among tree species in the Frisanco territory, the western part of the research area. Alder plantations are unknown in this area; data are included for old fields and former woodland.

chestnut groves, with small woods in which beech was frequently found. Trees had an important role in rural life. They were planted to provide nutrition for humans and animals as well as for other purposes.

Woody species can be subdivided into various categories according to the attitude of humans towards them in the past. Coniferous plantations made on abandoned farmland since the 1950s will not be taken into account. There is a large group of naturally regenerated species, growing frequently along roads, hedges, stone walls and piles of stones built in meadows. According to the farmers, there was no particular method of cultivating them since natural regeneration was easy. Among these species a large group was favoured by soil cultivation and associated activities such as the redistribution of stony material and fencing. The main produce from these trees was fodder but, incidentally, also litter, fuelwood and wood for various country crafts.

This group includes ash, maple, manna ash (*Fraxinus ornus*), elm (*Ulmus glabra*, *Ulmus campestris*), Lime (*Tilia cordata*), hornbeam (*Carpinus betulus*), hop-hornbeam (*Ostrya carpinifolia*), hazelnut (*Corylus avellana*), oaks (*Quercus petraea*, *Q. robur*, *Q. pubescens*), aspen (*Populus tremula*), black poplar (*Populus nigra*) and alder. Beech, growing only in small coppices, was mainly used for fuel but occasionally also for fodder. Cultivation activities did not favour its spread into fields and meadows.

A second group of species had been introduced for seed and fruit production, both for human and animal consumption. Their leaves could occasionally also be used for fodder, and wood was obviously a useful byproduct. These species are: chestnut, walnut (*Juglans regia*), apple (*Malus sylvestris*), pear (*Pyrus pyraster*), cherry (*Prunus avium*), and plum (*Prunus domestica*).

Other species were planted mainly for special purposes, such as: *Morus alba* for silkworm nutrition, and occasionally for fodder; *Robinia pseudoacacia* for pole production, erosion control and fodder; *Laburnum anagyroides* locally for hedges marking property boundaries; and *Alnus glutinosa* for fertilization.

A fourth group of species, supposedly quite small, includes those whose growth was hindered since they did not provide fruit, seed or fodder that could be used by humans or animals. These species were presumably eradicated or cut back. Old farmers usually group them in the category of 'shrubs' when interviewed.

Alder Plantations

Over a large part of the area it was common practice to plant alder in meadows in order to improve grass production (Fig. 5.3). Farmers assert that the litter of this species breaks down quickly and therefore fertilizes the soil. The nitrogen-fixing ability of *Alnus glutinosa* is obviously unknown to them. Seedlings used for plantations were collected where they appeared naturally in burned or

Fig. 5.3. Alders growing in meadows which are still mown.

mineral soil. Sometimes side branches were pruned off the bottom 2.0–2.5 m of the stem and multistemmed trees were occasionally thinned to only one stem.

According to the farmers we spoke to, sites chosen for alder plantations were those unfavourable for hay production with poor and superficial soils, exposed to the wind, or with southern slopes. Individual landowners sometimes planted areas a few hectares in area, but smaller plantations were more common.

Alder also provided small leafy twigs, which were collected in September, chopped into small fragments and mixed with manure in order to increase the amount of fertilizing material to spread on the fields and meadows in the following spring. Leaves were occasionally used also as fodder for pigs.

The origins of this practice are unknown. Nevertheless preliminary enquiries for the Land Register of 1826 mention plantation of alder and willow (*Salix* spp.) in the territory of Platischis in order to improve soil fertility (see Chapter 4). Alder grows naturally along creeks and wet sites in the region, but it seems it was virtually absent in most of the hills. The planting of alder ended during the early 1950s and only old farmers remember this technique. Today stem-pruning and shoot-thinning are practised only rarely since most of the fields and meadows have been abandoned.

Within the area of abundant alder distribution some smaller areas were identified where this species is absent and farmers ignore its use as fertilizer. Further investigation has shown that two of these districts have limestone as the soil parent material. This fact fits well with alder ecology, but also shows that land-management practices may be strictly bounded by environmental conditions.

Field observations show that alder is now distributed and abundant over a large area between the Judrio and the Torre valleys and grows on sites unusual to this species; there are no watercourses in the vicinity nor is the water table near the surface. Soils are frequently thin (10–20 cm deep). When growing on or near a mountain ridge the trees are heavily exposed to wind. Apparently alder seedlings were planted usually at random, but sometimes trees are clearly distributed at regular intervals of 4–5 m. In some cases single-stemmed trees are still recognizable, in other cases multistemmed stumps can be seen, either as a result of coppicing or as a result of thinning.

Alder forms both pure and mixed stands together with various hardwoods. The origin of these stands is not always quite clear, since alder does not regenerate easily by seed in abandoned meadows and fields, unless top soil and grass cover are removed locally. Young plants can be found at a distance of 4–5 m from the stump, around the crown area of coppiced trees. Seedlings can also be found under tree cover where succession has already created a full-density stand.

We examined alder trees growing in abandoned meadows not yet colonized by other species. Height growth is related to site conditions: trees growing on or near the ridge of the mountains are 2–3 m high, whereas in the lower part of the slopes they can reach 12–15 m. Exceptionally, alder reaches a diameter of 40–45 cm; the maximum age we have measured is approximately 60 years old. In thick mixed stands, alder shows a high mortality of both individual stems and trees growing under cover.

Under pure stands or even under individual trees isolated in abandoned meadows, the ground is sometimes covered with thick bramble (*Rubus* spp.) vegetation. Mixed stands developed on meadows, where alder was previously cultivated, show in some cases a rather peculiar distribution of other species: the biggest ash stems are always found near alder stumps, as if the 'protection' of this species could guarantee better growing conditions. Ash seedlings that germinate near alder stumps are obviously protected from the scythe or the mowing machine, but perhaps also other site factors are more favourable to regeneration and growth.

Soil Preparation and Cultivation

Information on this point is extremely scanty but some facts are already evident. Immediately after land abandonment, woody species spread naturally mainly around stone heaps and stone walls of various kinds, which had been built to improve soil productivity and to control the movements of domestic animals. In these locations regeneration is easier because young plants cannot be damaged by the scythe or by the hoe, are better protected from browsing animals and, in addition, grass competition is not so strong. Stone walls also provide a habitat for small rodents and a roost for birds, which are possible agents of seed dispersal. Finally, trees growing on these sites do not compete with herbaceous vegetation and therefore were accepted, even welcomed, by farmers.

After the abandonment of the land, the crowns of existing trees develop freely and sometimes the shade they cast restricts the growth of seedling trees in the immediate area. In meadows, in the absence of better ownership boundary markers, mowers used to leave (and sometimes still do) a thin strip (10–15 cm wide) of uncut grass along property limits where tree seedlings can survive and establish themselves. This way of regenerating can be detected only where hay making is still practised (Fig. 5.4), unless the boundary is marked by some kind of

Fig. 5.4. Grass and saplings surviving along property boundaries in mown meadows.

fence (e.g. barbed wire, wire netting, or wooden posts). Fences can also be directly made of trees (beech, ash, hornbeam), pruned or twisted into shape if necessary. The mowing of meadows favours tree species capable of vegetative reproduction from root suckers: *Populus tremula* and *Robinia pseudoacacia* are active colonizers in some areas.

LEAF COLLECTION FOR FODDER

Leaf collection for fodder was once a widespread practice all over Europe (see, for example, Haeggstrom, 1988; 1990). Nearly all species growing in the study area, including most fruit tree species, were utilized for fodder collection. Only chestnut was totally excluded from this practice, although little use was made of beech and alder. According to our sources elder (*Sambucus nigra*) and hazel were also exploited.

Livestock browsed directly from the trees and shrubs growing scattered in the pastures and in the few woods, but fodder collection was more common. Young twigs and branches were cut and either the leaves were separated by hand (leftover branches were subsequently used as fuel) or entire branches were tied together in bundles. Leaves were used as a green forage for all domestic animals, while dried bundles represented the main nourishment for sheep and goats during the winter.

Twig and branch collection was practised by lopping the trees, without reaching the highest part of the crown (Fig. 5.5). The size of the crown was drastically reduced through lopping and herbaceous vegetation had no need to compete for light. The highest part of the crown, not reached by lopping, was able to produce a large amount of seed, which enabled tree species to colonize abandoned fields and meadows quickly. In some cases, young trees were coppiced at the base of the stem.

All the species used for fodder, and most of those exploited as crop trees, are now well represented in secondary woodlands. Tree populations are formed partially by individuals exploited in the past as crop trees (e.g. producing fruits, seeds or fodder), but also by new individuals. The most successful species has been ash.

CONCLUSIONS

Available information on trees growing in fields and meadows and on their exploitation is very general and vague for Friuli (Feruglio, 1905; Selan, 1906; Marinelli, 1912; Musoni, 1915; Osterman, 1940). According to published sources, at the beginning of the 20th century, ash and fruit trees (including chestnut) were very common along the edges of fields and scattered in the meadows. Ash and alder were used as supplementary fodder in years when hay production was insufficient.

Fig. 5.5. Lopped ash photographed in 1972.

No mention is made of special cultivation and exploitation practices, and alder plantations are not mentioned. Nevertheless, as discussed above, this technique has been known for a long time. Moreno (1990) describes the use of *Alnus glutinosa* and perhaps *A. incana* in Ligurian pastures at the end of the 18th century. The plentiful information on how the various woody species were used by farmers, which we have obtained through interviews, has been ignored by the agronomic and ethnographic literature.

Our interpretation of some important characteristics of secondary woodland structure and composition has been made possible by the knowledge of rural customary practices obtained through interviews. The most obvious is the distribution and the abundance of alder.

A remarkable feature is the wide distribution of *Fraxinus excelsior-Acer pseudoplatanus* woodland mixed with individuals of *Alnus glutinosa.* This species is represented by scattered trees, in many cases included in the dominant layer, comparatively old and sometimes already dying back. Moreover, in some cases, the ash distribution coincides with that of alder (Fig. 5.6); this is a result of the mechanical protection of young individuals of ash.

It can be supposed that alder has also modified site conditions and, as a consequence, interfered in regeneration, growth and competition of other species whose range could also have been influenced. The widespread distribution of ash-maple stands, even on sites that are not best suited to these species, could be due to the influence of alder on soil characteristics. At the same time the large quantities of ash and maple seed from the crowns of lopped trees may have accelerated the colonization of abandoned fields. This fact makes the interpretation of the present situation more difficult.

Human influence on natural succession has been exerted both directly, by favouring or hindering a certain species, and indirectly, through the modification

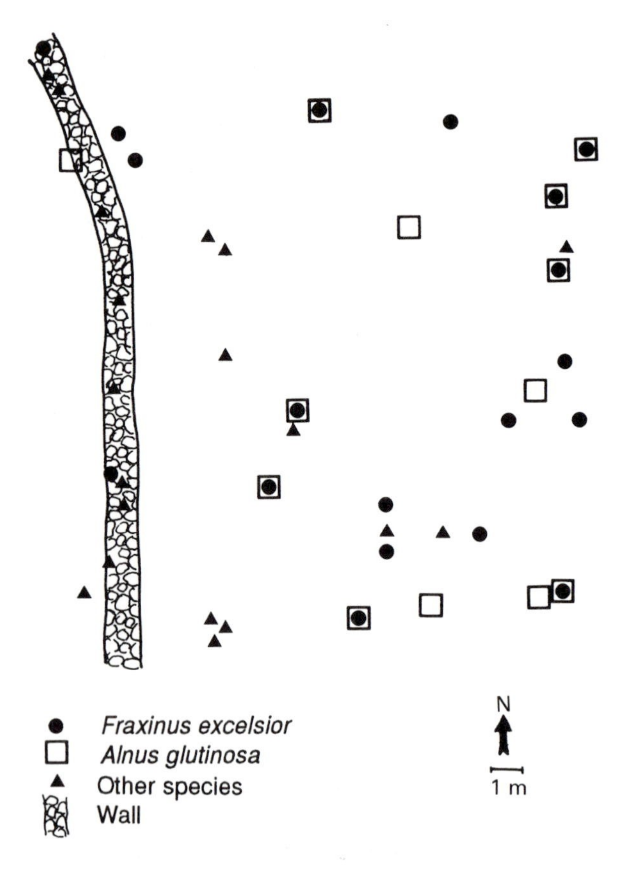

Fig. 5.6. The distribution of alder and ash in an abandoned meadow. Ash grows mainly close to alder stumps and stone walls.

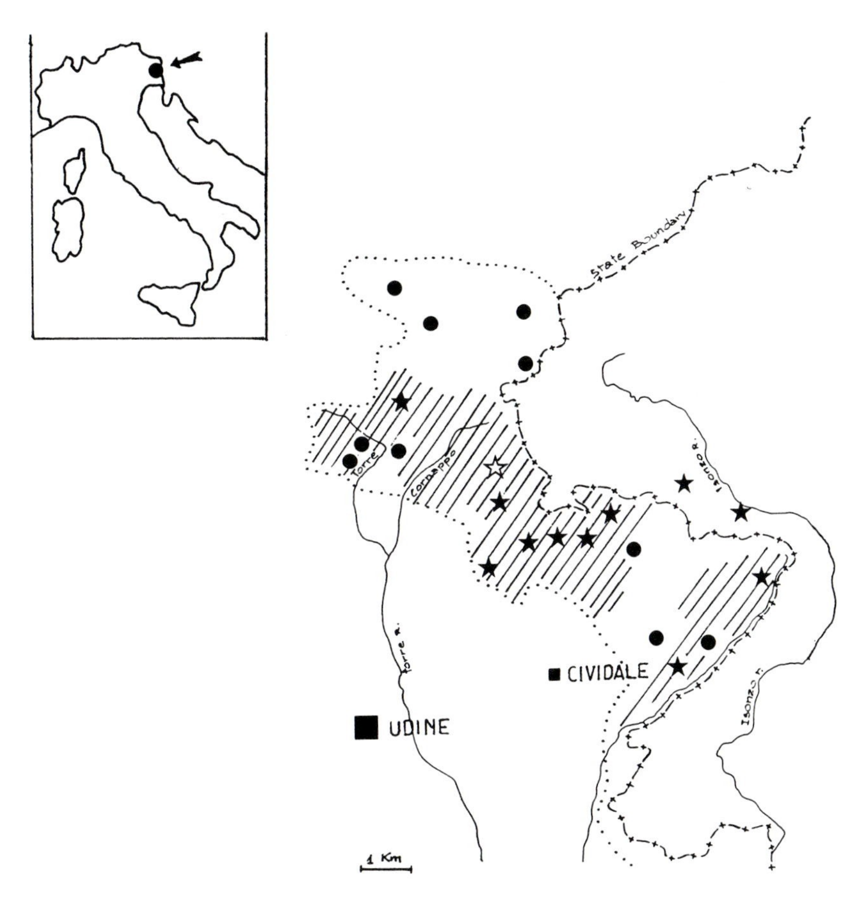

Fig. 5.7. This maps shows areas where alder was planted in the eastern part of the study area. Full stars show the knowledge of the practice derived from interviews. Full dots show ignorance of the practice. Dotted lines show the borders of the Slavic-speaking population at the beginning of the century. Diagonal lines show the approximate distribution of mixed stands with alder. Empty stars show places mentioned in the Land Register of 1826.

of the physical environment. Both kinds of action have been, in turn, controlled by the way natural resources were managed for animal husbandry and agriculture. Activities were concerned with individual trees and not forest stands; one cannot therefore talk about silvicultural practices. Within the study area some techniques were practised everywhere, while others had a narrower range. The range of alder cultivation, according to our interviewees, is congruent with the area of Slavic population. We do not as yet have a satisfying explanation for this coincidence (Fig. 5.7). It is clear, however, that vegetation dynamics and the characteristics of the new forest landscape are partially determined by past rural activity.

6

Deforestation and Natural Regeneration of Woodland: The Forest History of Molise, Italy, over the Last Two Centuries

P. Di Martino

Summary

The history of woodlands in Molise is closely linked with the history of pastoral farming. A specialized form of woodland management had developed alongside the system of transhumance and grazing, which was in place until the last few years of the 19th century. From the beginning of the 20th century the abandonment of transhumance practices in the coastal and hilly areas of Molise resulted in the devastation of woodlands and their conversion to agricultural land. Over the same period, there has been a slow and gradual spread of woodland by natural regeneration over the abandoned pastures and arable fields of the more mountainous parts of Molise.

Introduction

The landscape of southern Italy changed dramatically during the 19th century. In this chapter, the factors affecting woodland change in this period are studied by making a close examination of the landscape of Molise. This region is the smallest region in the south of Italy and it is therefore possible to carry out a detailed analysis of relatively small-scale changes in land use (Fig. 6.1). Over the period in question, deforestation was certainly the most evident cause of forest change and its devastating impacts on the landscape are familiar from many studies of forest history. However, there is considerably less information about what was happening in surviving woodland. The ecology of these woods, including the mixture of species and woodland structure, has been strongly influenced by human activities. Over the past two centuries the intensity of these activities, and the nature of the impacts, has varied depending upon a wide range of social and

Fig. 6.1. The location of Molise.

economic factors. This chapter presents results of detailed investigations into the ecological history of the forests of Molise and provides information on the factors that helped to maintain its woods in the past.

SOURCES OF INFORMATION ON FOREST HISTORY

The main sources used to reconstruct past forest landscapes and land use are censuses and statistics including cadastral documents from 1816 onwards, agricultural censuses, the national forest inventory and other miscellaneous statistics. As usual, these sources present some problems, such as the reliability of records and especially the compatibility of information from different periods: the categories of land use and the typology used in recording areas are inconsistent; in addition the boundaries of Molise changed from time to time. These problems have been resolved partially by analysing and comparing data of different periods for every single commune in terms of percentage of the total land area; by grouping municipal data district by district and by using the oldest boundaries of the region as the reference area.

A particularly valuable group of sources is made up of archive maps of woods including the important first maps of the Italian Geographical Military Institute produced between 1869 and 1875. These maps can be compared with more recent cartography and aerial photographs to obtain a knowledge of landscape changes at particular sites. This method partially makes up for the lack of modern sources for forestry statistics and cartography. The regional forestry administration has no forestry maps or inventory and this makes it difficult accurately to position woods and plantations. It is therefore impossible to assess forest landscape changes at the regional scale.

Archive documents provide a rich source of information for 19th century woodlands. A special forestry inventory known as the *Verbali di Verificazione*, which was developed from 1840 to 1859, provides a good range of data including information on the location of woodland, its extent, species mix, stand structure, the way it was managed and its use.

THE STUDY AREA

Molise covers an area of 4400 km^2 and is mostly mountainous with elevations up to almost 2000 m above sea level. In the interior there is an irregular landscape formed by high hills and small valleys. The Adriatic coast is only a few kilometres in length. The climate is non-uniform: temperature decreases from the hilly coastal region to the interior where cold winters with heavy snow are experienced. The average temperature of the warmest month is higher than 20°C. Precipitation is generally at its highest in November and its lowest in July.

The settlement pattern is very dispersed – especially in the Apennines. The average population density in 1981 was 74 inhabitants/km^2. The topography and geography of the region provide unfavourable conditions for agriculture. Nevertheless, Molise has the highest proportion of agricultural land compared with other Italian regions (Fig. 6.2). Woodland covers 70 561 ha with the main species being *Quercus pubescens*, *Quercus cerris*, *Abies alba* and *Fagus sylvatica*. The majority of the woodland consists of coppice and the larger share of property is owned by communes (Fig. 6.3).

FOREST MANAGEMENT IN THE FIRST HALF OF THE 19TH CENTURY

The forest history of the 19th century is closely linked to the social and political events that occurred in southern Italy during the early part of the century, particularly the abolition of the feudal system during the period of French government. All municipal properties that included feudal land over which ancient uses were exercised were subdivided between citizens. In the period following subdivision, archival evidence points to an enormous amount of

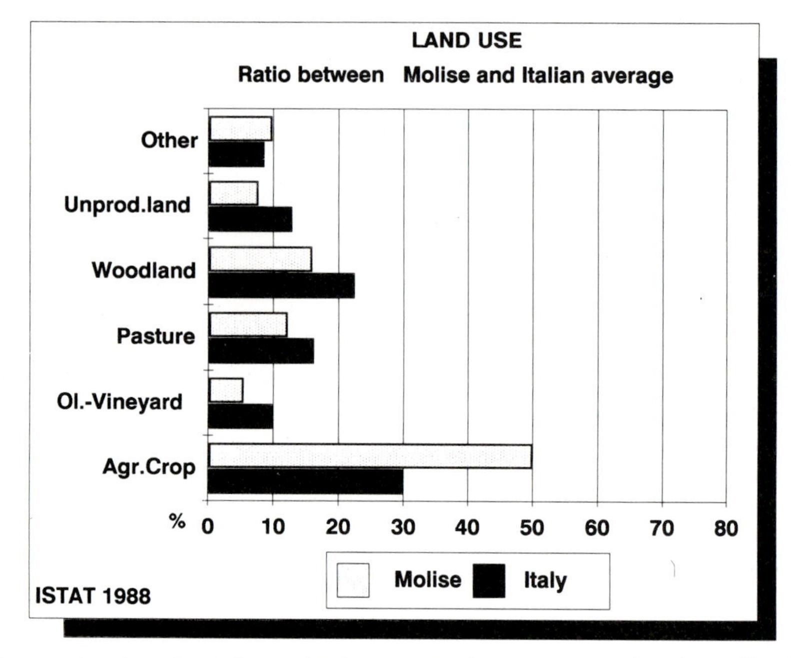

Fig. 6.2. Land use in Molise and Italy in 1988. There are 70 561 ha of woodland in Molise which is 15.9% of the land area. The Italian average is 22.5%. (Source: ISTAT, 1988.)

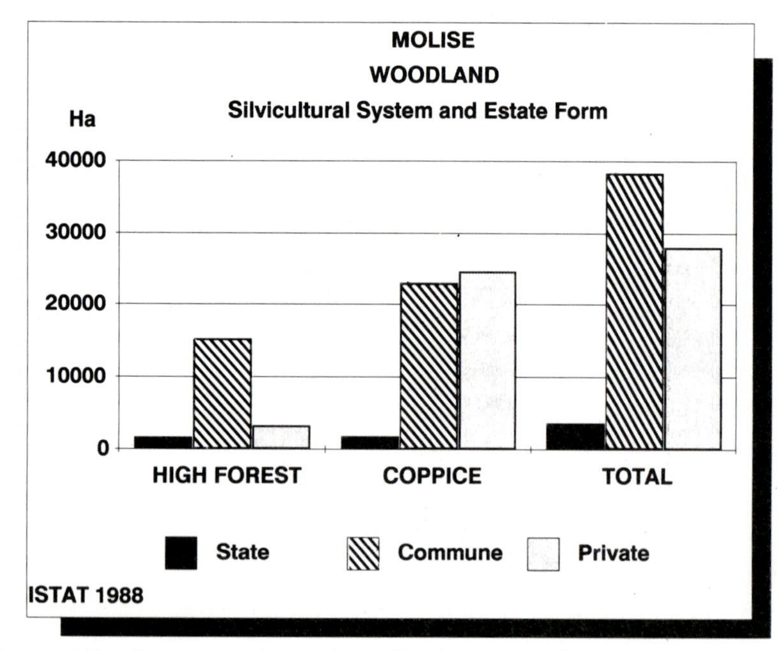

Fig. 6.3. Woodland ownership and woodland type in Molise, 1988.

woodland being cleared and converted to agricultural use. It is, however, very difficult to estimate overall land-use changes owing to the lack of contemporary regional statistics.

According to the available statistics, the area of woodland in the region showed little change between 1816 and 1877; that is, from the establishment of the cadastral system to the introduction of the post-unification forestry law. In 1836, there were 81 000 ha of woodland in Molise (Del Re, 1836). Woodland was distributed principally between communes (54%) and private owners (41%); only 5% belonged to the state or corporations. Data from 1836 (Fig. 6.4) show over half the land area as arable. This was the result of the massive deforestation and great expansion of cereal cultivation during the second half of the 18th century and the first decade of the 19th century. At this period, there was a great increase of population that, according to Massafra (1980), 'makes to double the population of Molise that, except to create new needs, causes an excess of labour that can only find employment in agriculture'.

The 1836 data can be used to show land-use variations between the three ancient districts of Molise: Isernia, Campobasso and Larino (Fig. 6.5). Campobasso had the smallest areas of pasture and woodland and the largest areas of arable, olive yards and vineyards. Isernia, the western and most mountainous district, had the smallest amount of arable land and the most pasture, while Larino in the east had most woodland.

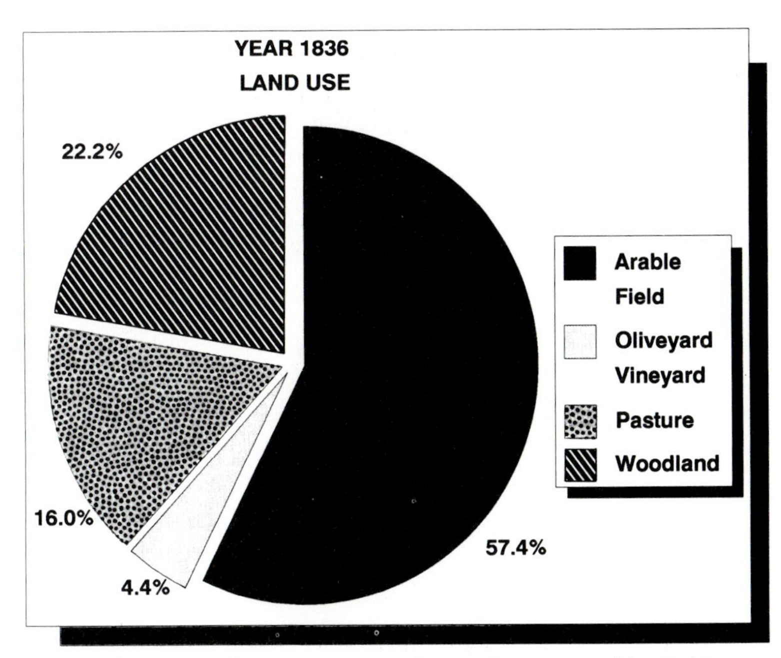

Fig. 6.4. Land use in Molise in 1836 according to data reported by Del Re.

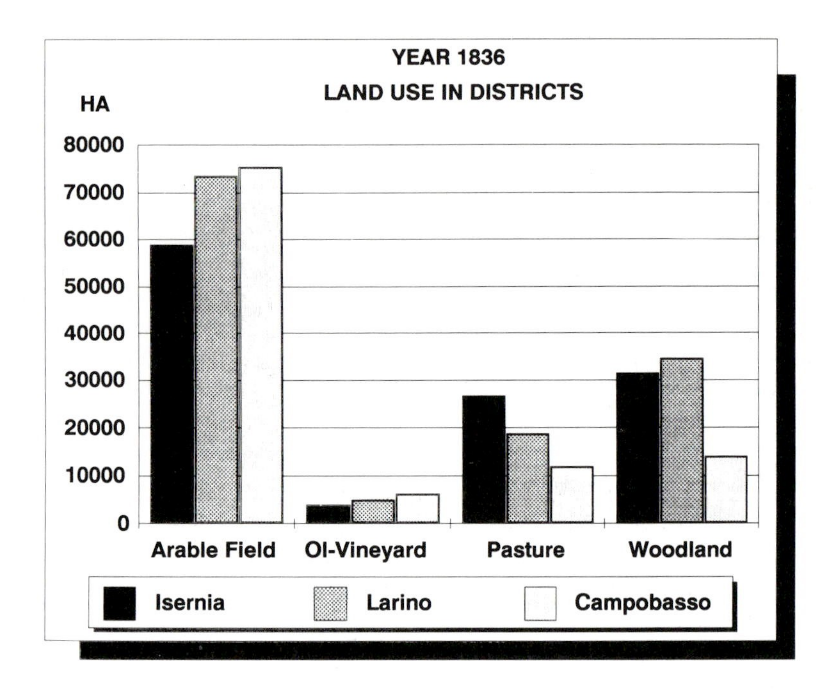

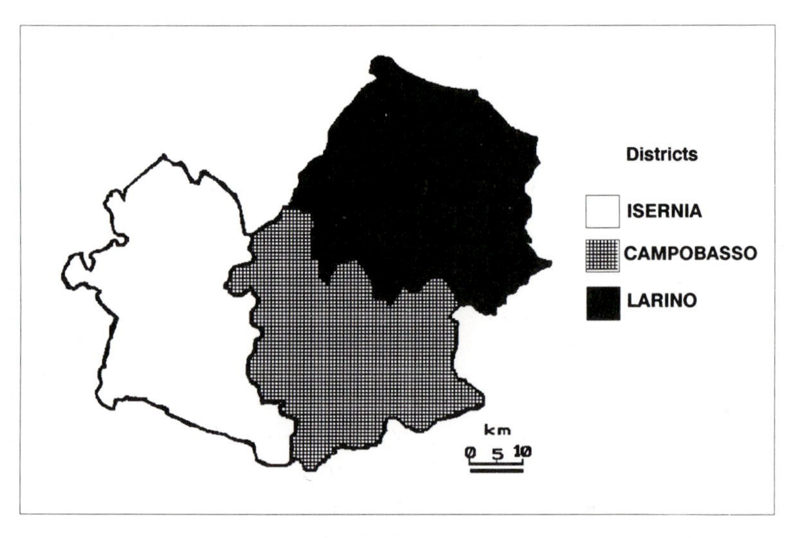

Fig. 6.5. The ancient districts of Molise and their land use in 1836.

The basis of the forest legal system was the Forestry Law of 1826, which represented the first attempt at modifying woodland management in southern Italy. It included new instructions about geological survey and regulations on the exploitation of municipal woods where private owners were excluded. In these woods, the legislation specified a *regular cut* that was a kind of clear felling with at least 45 trees/ha being left standing. However, the law did not establish instructions to improve or to conserve woods. According to Pepe (1811; 1834) there was a lack of silvicultural technique and the woodlands were exploited irregularly and under pressure from grazing and agricultural uses. He considered that the principal factors affecting woodland density and structure were clearance, usurpation, pollarding and damage to natural regeneration.

Only 12 years later, in 1838, records began to be collected for the first forestry census. In Molise the census was completed in 1840, but 4 years later the regional administration issued a 'sovereign decree for the improvement and restoration of reproduction of the municipal woods' and instituted a new forest survey. I have found these survey records for 143 municipal forests with a total area of 25 000 ha from the period between 1840 and 1859. These provide information of high quality and enable us to understand the conditions of forests in Molise in the mid-19th century.

The woods varied in size from 6 to 1658 ha. In the coastal area, the dominant species were *Quercus pubescens* and *Quercus cerris. Quercus ilex, Carpinus betulus, Fraxinus ornus* and *Pistacia lentiscus* were associated species. Inland, the principal species were *Quercus cerris, Fagus sylvatica* and *Abies alba.* (I have stored the information from the survey in a database in order to enable full analysis.) The area of woodland was recorded in two different ways: total area and woodland area. The total area covered by the records was 28 539 ha while the woodland area was 24 860 ha. The difference was made up as follows: clearings (2119 ha); agricultural fields (1464 ha); landslides (74 ha) and burnt areas (22 ha).

Figures 6.6, 6.7 and 6.8 show the irregularity of the stand structure and density of the woodlands. More than half of the woodland area contained a scatter of old trees (woodland pasture) and documents often show an irregular distribution of trees and great variation in the amount of coppice and young trees. Of the area surveyed, 53% included trees with diameters greater than 45 cm, while just under a quarter (24.4%) had trees with diameters greater than 65 cm (including some trees 120 cm in diameter). Woodland found at 700 m above sea level or higher had the greatest density of trees.

The data suggest that stand structure was generally composed of:

- a multi-storey, consisting of an overwood of a few old trees, many of which had been pollarded or lopped;
- an intermediate storey composed of old coppice shoots;
- an underwood with shrub coppice and a small amount of natural regeneration.

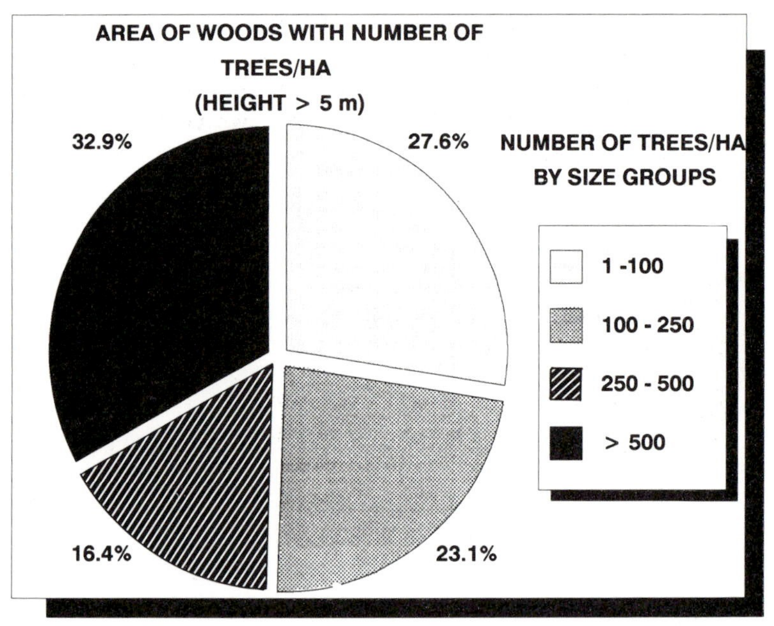

Fig. 6.6. Percentage area of woodland with different densities of trees greater than 5 m in height. Summary of data for 143 woods from the *Verbali di Verificazione* 1840–1859.

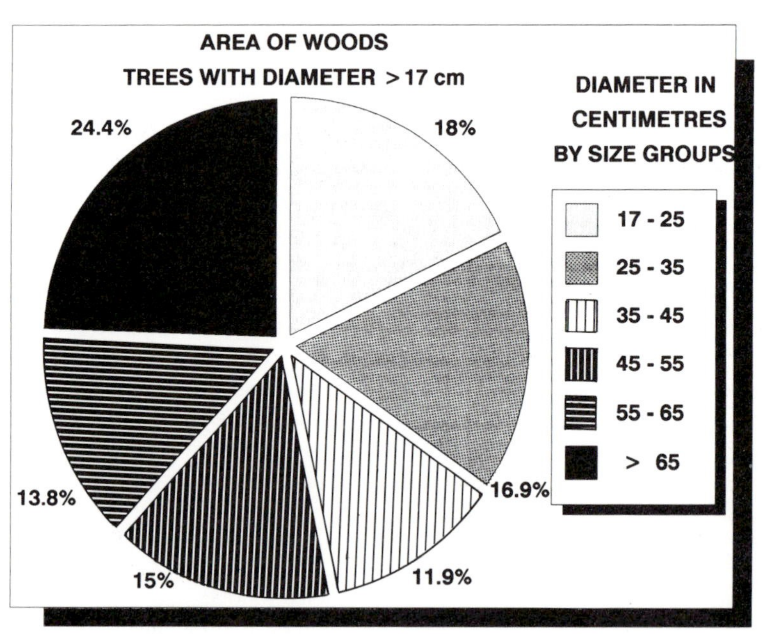

Fig. 6.7. Percentage area of woodland with trees of different diameters. Summary of data for 143 woods from the *Verbali di Verificazione* 1840–1859.

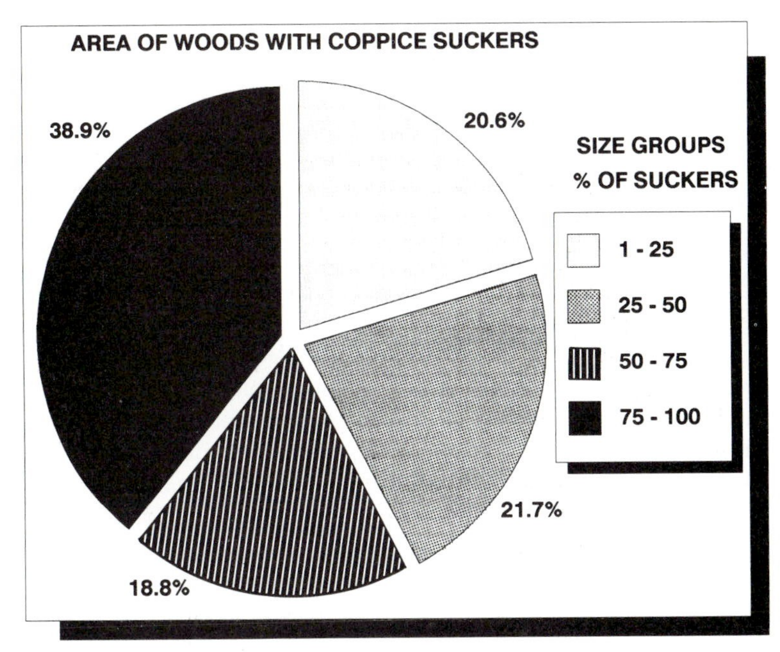

Fig. 6.8. Percentage area of woodland with different proportions of coppice suckers. Summary of data for 143 woods from the *Verbali di Verificazione* 1840–1859.

The stand structure varied depending on the location of the wood. Small coppiced areas were found in more accessible places – especially near paths – while areas covered by shrub coppice and old overgrown standard trees were generally more distant from settlements.

There is no doubt, therefore, that in spite of the regulations of the 1826 forestry law, illegal exploitation was dominant and the forestry administration's attempt to order regular cuts in sections for municipal woodland had failed. Granata (1839) pointed out that the type of management proposed in the 1826 forestry law proved contrary to the interest of the owners because the imposed cut rate was too high. Moreover, the communes required woodland products for charcoal production and firewood rather than high-quality timber. In addition, grazing of the woodlands continued because of the importance of earnings from the rent charges for wood pasture in the local economy.

According to Pepe (1844) 'the woods situated in internal territory and in the west and mountain areas of Molise are so far from the sea, the main road and towns, that the exploitation of wood is not remunerative'. Documentary evidence, for example, shows that in a section of the municipal wood Collerotondo, in Montagano territory, it was proposed to cut 330 old Turkey oaks with an estimated production of 255 tons of firewood. The commune had great difficulties in finding purchasers because of the technological difficulties of felling the trees and transporting the produce. As a result, the proposal was abandoned. In some cases, however, there is documentary evidence of regular cuts in sections of coppices on the mountain sides, especially in the Matese mountains that today form the boundary between Molise and Campania. Clearly, there was a more active market for wood products in this part of the region.

The management of Montedimezzo Wood, which was part of a Royal estate and of one of the most important agro-pastoral farms in southern Italy (linked by the transhumance system to a large estate in Apulia from the 17th century) was unique. From 1845 to 1855 regular cuts were made in sections and the fellings were massive. The effects of this mid-19th century exploitation are still visible in the current floristic composition of the forest (Di Martino, 1986).

The value given of woodland in contemporary documents was often directly related to its use as wood pasture and this fact often had a strong effect on the management of woods. For many private and municipal woodland owners in Molise, renting out woodland pasture produced significant income. Pepe (1844) provides confirmation: 'In our rural economy of Molise the woodlands produce two pastures during the two seasons of summer and winter and we divide pastures into two seasonal renting. If a wood is situated in the mountains, with springs and fertile soil, summer grass has more value and with our migratory stock-raising it is easy to rent. Winter pasture is much sought after when wood does not produce acorns because soil is not damaged by swine grazing'. Regarding the improvement of wood pasture, Pepe considered that underwood and shrubs together with old trees and pollards could be cut to increase the area of pasture.

The documents provide much information on projects to improve woodland such as: (i) the cutting of old and pollarded trees (generally oaks); (ii) the elimination of thorny shrubs; (iii) the cutting of scrub coppice down to the root; (iv) the thinning of mature coppice; (v) the sowing of acorns in clearings; and (vi) the enclosure of felled and reafforested areas by the construction of hedges made from cut thorny shrub or dense coppice. One detailed method of tree establishment is shown by records of 1839 for Frosolone, which show that the Regional Forest Authority planted three acorns in single holes at a distance of 1 m. However, as discussed earlier, in practice many obstacles were placed in the way of this woodland 'improvement'. In 1839 the regional forestry administration afforested about 50 sites on clearings and on eroded soil. These were all subsequently destroyed by illegal grazing.

The Role of Transhumance

The seasonal grazing associated with transhumance played an important role in woodland used for animal forage. The regional form of transhumance was based on an ancient economic and geographical organization of pastoralism sanctioned by Alfonso d'Aragona in 1447. It enabled the unified management of stock pasture during the seasonal migration of animals from the great Apulian plain to the Abruzzi mountains along the routes of transhumance called *tratturi*, which spread over the whole of Molise (Fig. 6.9).

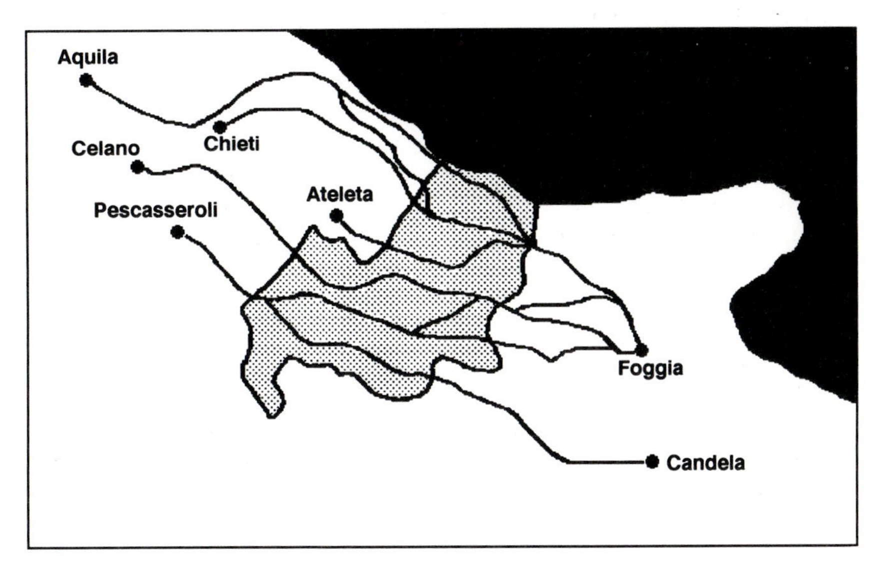

Fig. 6.9. The network of *tratturi* in Molise, Abruzzi, Campania, Basilicata and Puglia. There were 398 km of *tratturi* in Molise covering an area of 3600 ha.

There is little information on the numbers of animals passing through Molise, but Sprengel (1971) has made some estimates (Table 6.1). It is clear that from May to September thousands of migratory livestock, mostly sheep with goats and cattle, grazed in the forests along the transhumance routes. In addition, there was grazing by permanent livestock. The effects of this grazing on the woodland can be summarized as follows:

Table 6.1. Estimate of the number of animals involved in transhumance at different dates.

Year	Livestock in transhumance
1815	950 000
1840	1 200 000
1860	760 000
1877	730 000
1951	400 000

Source: Sprengel, 1971.

- damage to the natural regeneration, directly by feeding and also by trampling;
- damage to young coppice areas;
- change to the stand structure and canopy density so to favour the production of grass and, in the case of swine pasture, the production of acorns;
- selection and conservation of species for the production of fruits (*Pyrus pyraster, Cornus mas, Cornus sanguinea*);
- pollarding and lopping;
- felling of trees for the construction of sheds and fences and cutting of thorny shrubs (*Crataegus, Prunus, Rubus*) to make temporary enclosures.

An interesting example of these practices is provided by the municipal wood of Rotello (Fig. 6.10). Documents state the number of livestock grazing in this wood pasture. In 1849 there were 469 cattle, 169 horses, 2700 sheep, 700 goats and 730 swine grazing on a woodland area of 302 ha. Two further examples will be given to show the range of information available for individual wood pastures. In 1847 Difesa Wood, Colletorto (270 m above sea level), comprised 762 ha.

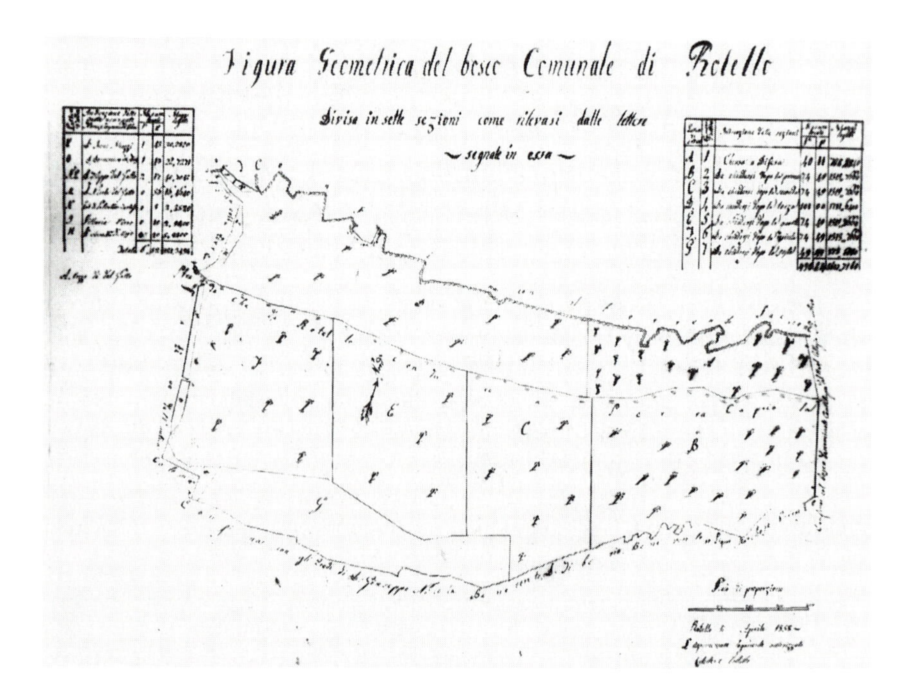

Fig. 6.10. Map of the municipal Rotello Wood drawn in 1858. The forest administration was attempting at this time to draw up a plan to fell the wood in sections. Ancient documents describe this wood as a wood pasture with scattered old trees, usually pollarded oaks (see enlargement from map opposite).

There was a scatter of old trees, which were pollarded and lopped, and coppice. The trees were damaged by 'abusive cuts' and grazing. It was planned to eliminate the old trees, and to cut the coppice down to the root. Table 6.2 shows the size distribution of oak trees in this wood. At Difesa Wood, Roccasicura, the records of 1846 showed that the 87 ha wood (800 m above sea level) had been periodically exploited from 1825 to 1841. This exploitation had taken the form of the thinning of young trees to reduce the stand canopy and to favour grass production for grazing, excluding old trees, which were retained for the production of acorns and pollarding (Table 6.3). These examples confirm that as regards the modification of stand structure, different practices were carried out depending on altitude, topography and the abundance of pasture.

After the unification of Italy and the enactment of the forestry law of 1877, much of Molise – and especially the hilly coastal area – was subjected to massive deforestation. In private woods and municipal property (*quotizzazioni*) deforestation followed directly upon the process of land reform and the division of

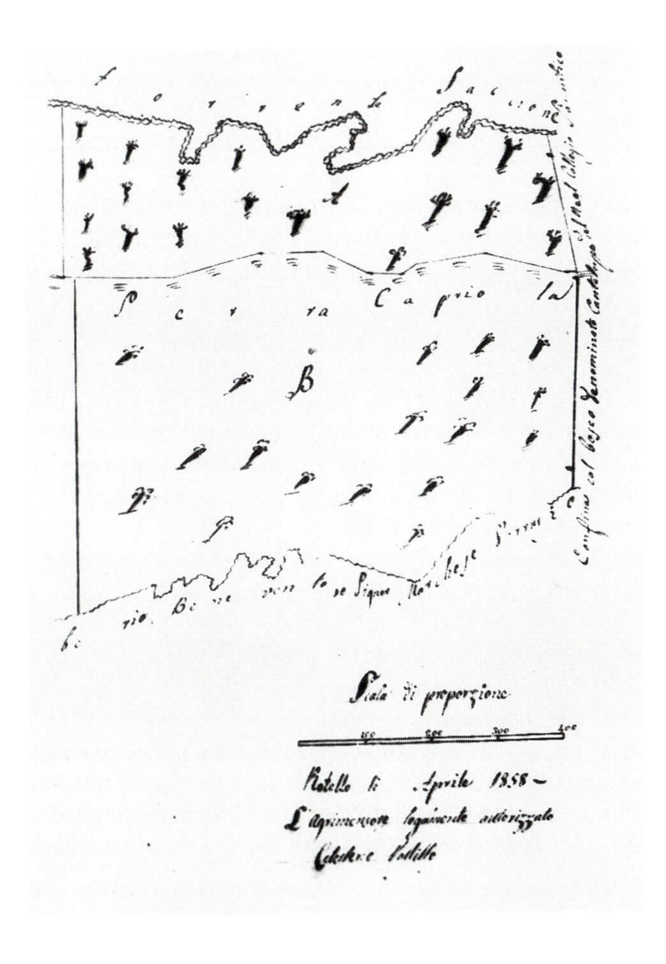

Table 6.2. Height, diameter and density of oak trees in Difesa Wood, Colletorto, in 1847.

Species	Height (m)	Diameter (cm)	No. of trees/ha
Quercus pubescens	13	29	10
Quercus pubescens	14	22	9
Quercus pubescens	3	12	435
Quercus cerris	13	29	1
Quercus cerris	14	22	14
Quercus cerris	3	12	258

Table 6.3. Height, diameter and density of oak trees in Difesa Wood, Roccasicura, in 1846.

Species	Height (m)	Diameter (cm)	No. of trees/ha
Quercus cerris	14	65	209
Quercus cerris	12	45	200
Quercus cerris	3	17	117

property. Archive documents show that from 1880 to 1895 about 15 000 ha of woodland was cleared for agricultural use. The history of Ramitello Wood is a good example. This was a private wood of 2489 ha situated on the coast near the Puglian boundary (Fig. 6.11). The species and stand structure for 1840 are given in Table 6.4. At that date it was a mist oak wood with a great dominance of *Quercus cerris*. It was used as wood pasture and would have been grazed by stock migrating from nearby Puglia. From 1880 onwards, the forest was gradually cleared for agricultural use and this devastation continued in the 20th century as part of the scheme of land reclamation for the whole area (Fig. 6.12). A few hectares of this woodland, consisting of *Quercus cerris* and *Quercus ilex*, survive on the hills near the coast (Taffetani, 1991).

Angeloni (1885) noted that more than 14 000 ha of woodland were not legally protected from clearance. In the remaining protected woods, the forestry commission continued to fell regularly as established by the older 1826 forestry law. According to Bellini (1879), 'in the regular cut in sections as well in the total area, at least one or two trees every 10 m had to be left standing'. During the same period the transhumance system began to decline. The 1877 forestry law meant that grazing was restricted as there was more surveillance in the woods; ancient practices were modified. Some Apennine grassland was gradually transformed by agro-pastoral exploitation. Today there is still evidence of this activity in the form of large enclosures with sheds of varying shapes made of drystone walls (Fig. 6.13).

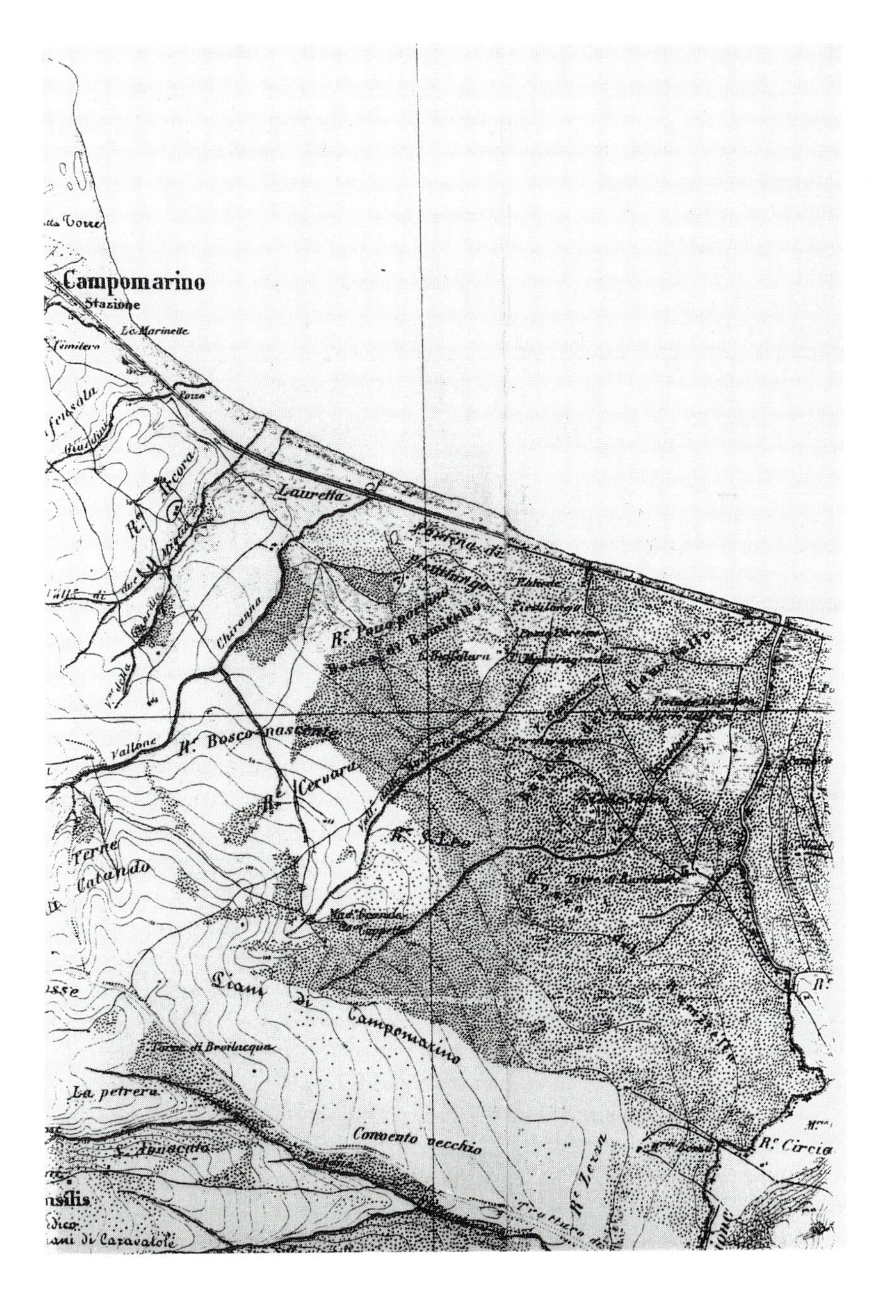

Fig. 6.11. Detail from a map of 1870 (1:50000) which includes part of the coast of Molise and a large area covered by Ramitello Wood extending to 2000 ha. This is an example of one of the many woods once found along the coast. (*Dai tipi dell'Istituto Geografico Militare,* Autorizzazione n. 3562 del 21/02/1992.)

Table 6.4. Height, diameter and density of trees of different species in Ramitello Wood, 1840.

Species	Height (m)	Diameter (cm)	No. of trees/ha
Quercus cerris	11–16	34–67	108
Quercus cerris	8–11	25–34	1
Quercus pubescens	11–13	42–84	2
Quercus pubescens	8–9	17–25	2
Quercus ilex	7–8	34–84	3
Carpinus betulus	—	4–17	6
Cornus mas	—	8–17	4
Pyrus pyraster	—	—	1
Fraxinus ornus	—	4–17	—
Populus spp.	—	—	2

The surviving woods were increasingly exploited because the market for wood products improved and the transport network became more established. Examples of intensive felling are provided by the history of three silver fir woods, Abeti Soprani, Vallazzuna and Montecastelbarone (Di Martino, 1988). Taken together these woods covered about 500 ha and from 1900 to 1919 a total of 98 000 m^3 of fir wood were cut. Today, in spite of this scale of felling, silver fir remains and is invading neighbouring areas of uneven-aged beech coppice with standards. In the larger woods felling plans were designed, and from 1920 onwards coppices began to be converted into high forest.

The large-scale emigration from Molise at the beginning of this century had enormous impact on the rural landscape as it resulted in land abandonment and extensive land-use change – especially in the Apennines. Factors influencing the exodus from the land included low earnings from silvo-pastoral activities and cereal cultivation, the scattered location of fields, the unfavourable characteristics of the soil, the lack of communication and the low level of industrialization. Today, 101 out of a total of 138 communes have a population of under 2000 inhabitants.

From 1930 onwards afforestation with *Pinus nigra* subspecies *italica* (Villetta Barrea), took place on 1000 ha on 110 different sites. Some plantations were made on limestone soils, and some were on devastated woods and abandoned pastures. According to Guidi (1985) those *Pinus nigra* plantations made on dry sites more suitable for oak are gradually developing into mixed woodland. After 1950, the species used in afforestation schemes was changed to include other species such as *Pinus halepensis, Pinus pinea, Eucalyptus globulus, Acacia saligna, Cupressus sempervirens* and *arizonica, Cedrus deodara* and *atlantica, Abies alba, Picea abies* and *Alnus cordata.* Post-war forestry plans affect 60 woods with a total area of 15 000 ha in 31 communes. Many of these plans

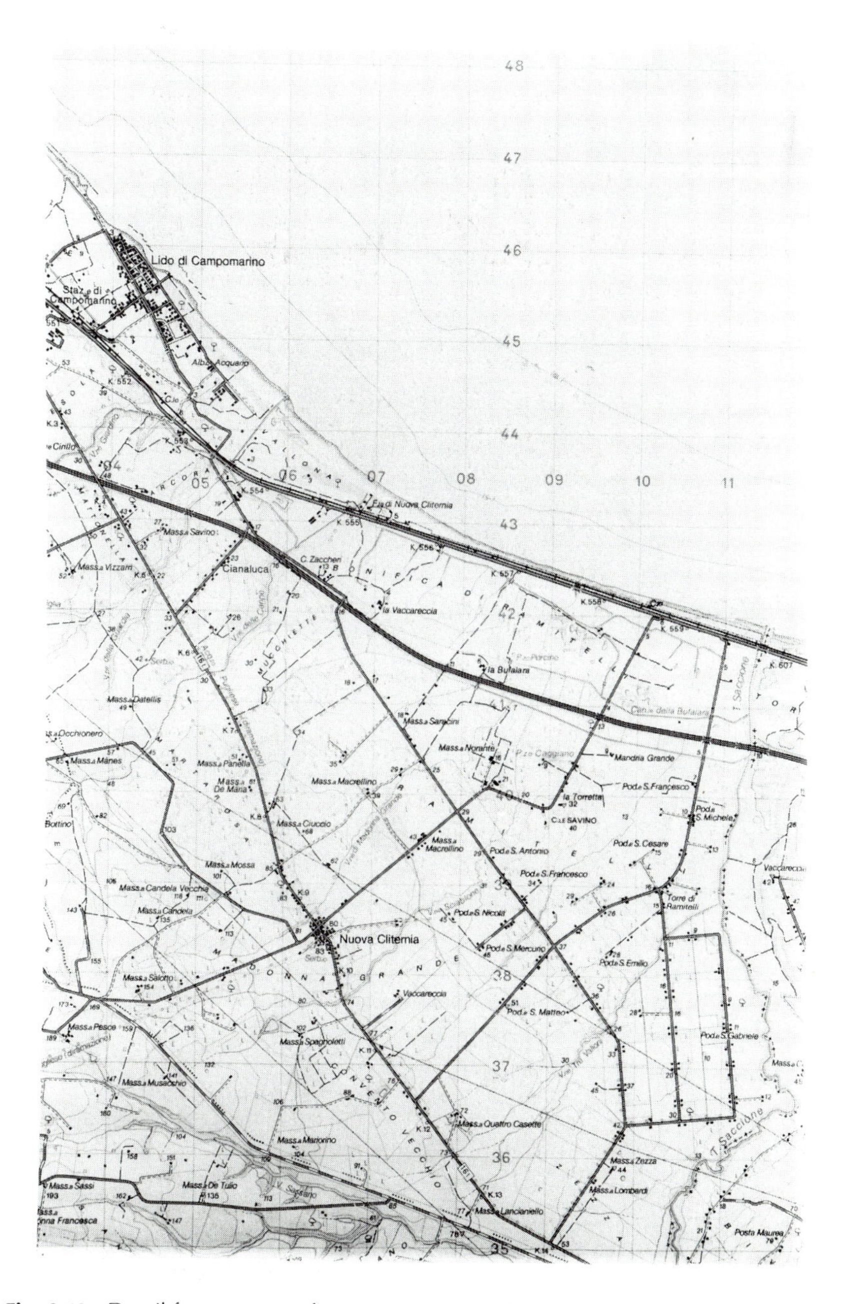

Fig. 6.12. Detail from a map of 1982 (1:50000) which includes the same area as Fig. 6.11). The few remaining woods consist of riparian woods along streams and small farm woods. (*Dai tipi dell'Istituto Geografico Militare,* Autorizzazione n. 3562 del 21/02/92.)

Fig. 6.13. Aerial photograph of the western part of the territory of Capracotta, about 1370 m above sea level, taken in 1986. Many enclosures with several sheds constructed of drystone walls are visible over an area of about 300 ha. (Authorization S.M.A. n.0/98.)

have yet to be reviewed. The remaining 55 000 ha, of woodland, mainly coppices, are managed according to felling plans.

Evidence for the Natural Regeneration of Woodland

The total forest area of 65 519 ha given by the Agricultural Census takes into account only productive forests. If this area is compared with the ISTAT data of 70 561 ha, about 5000 ha of woodland seems to have developed naturally on abandoned and unproductive fields (Pettenella, 1988). Data concerning farms that have wooded areas as part of their productive land suggest that there is a large number of farms (18 233 or 79.5%) between 1 ha and 10 ha in area with an average wood extension of 0.6 ha (Table 6.5).

Broad landscape changes for the three main regions of Molise are shown in Figures 6.14, 6.15 and 6.16. These figures should be interpreted with care. Data obtained from the 1929 cadastral documents concerning cultivated lands were compiled under fascism and probably include overestimates. Data from the 1982 Agricultural Census only include woodland on productive land and exclude, as

Table 6.5. Area of woodland on farms of different sizes in Molise in 1988.

Size groups (ha)	No. of farms	Woodland area (ha)	Average area (ha)
1	1157	256	0.2
1–2	2766	1003	0.4
2–5	7836	4542	0.6
5–9	6474	6031	0.9
10–20	3228	5088	1.6
20–50	1097	3738	3.4
>50	365	44861	122.9

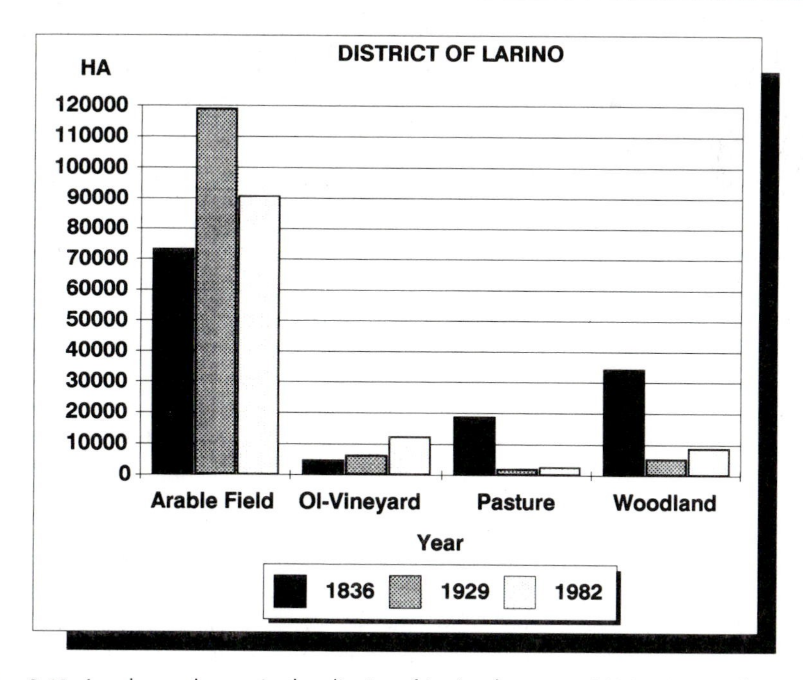

Fig. 6.14. Land use change in the district of Larino between 1836, 1929 and 1982.

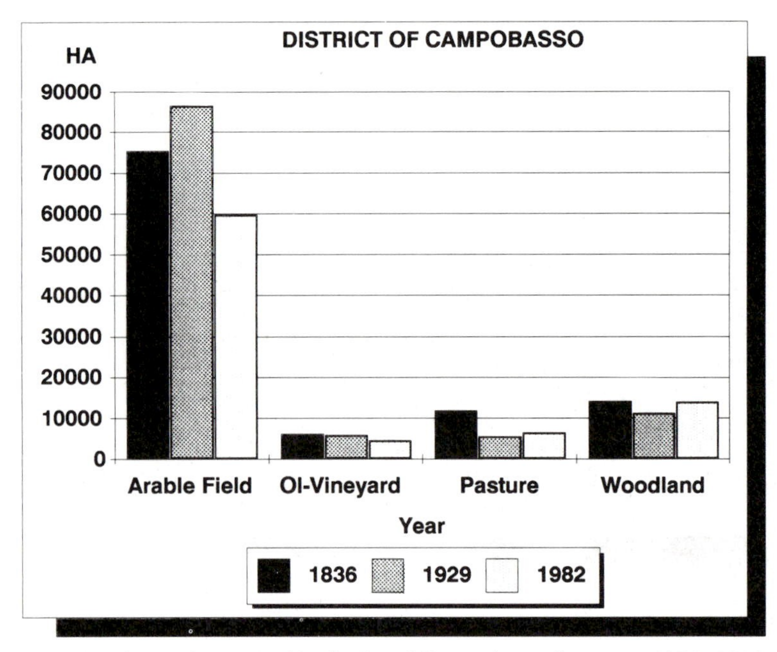

Fig. 6.15. Land use change in the district of Campobasso between 1836, 1929 and 1982.

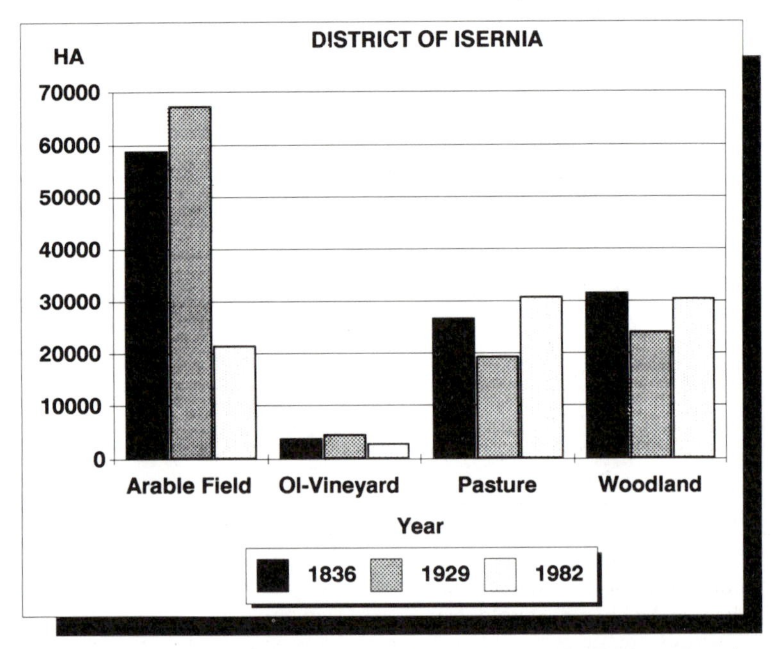

Fig. 6.16. Land use change in the district of Isernia between 1836, 1929 and 1982.

mentioned above, abandoned and unproductive fields. The charts confirm the great changes in land-use in the district of Larino to the detriment of woodland and pasture, while in the other two districts there is a large decrease in arable land – especially in the district of Isernia (from 59 158 ha in 1836 to 21 735 ha in 1982) – and an apparent balance as regards woodland. Figures 6.17, 6.18 and 6.19

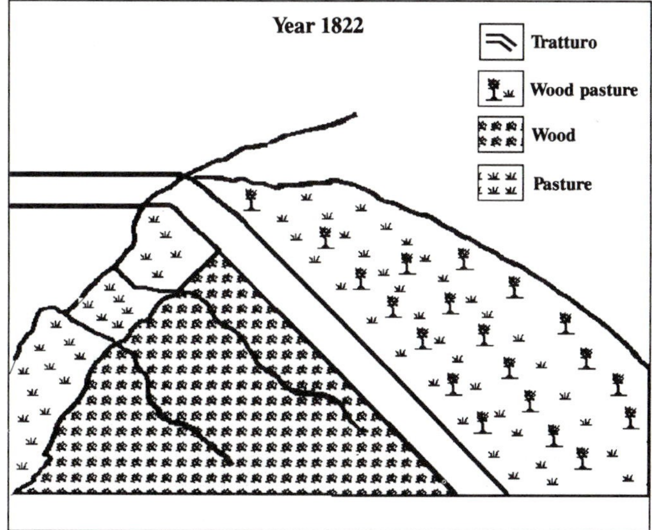

Fig. 6.17. Above: Aerial photograph of Montedimezzo Wood taken in 1986 (Authorization S.M.A. n.0/98). The *tratturo* Celano-Foggia is still visible and has a width of 111 m. Below: Map compiled by the author from a map made in 1822 1822 kept in Archivio di Stato di Napoli, Archivio amministrativo di Casa Reale, III inventario, Segreteria, fs 1116/1; f.50. The wood at the upper margin of the *tratturo* was a wood pasture while the rest of the forest coverage has remained virtually unchanged.

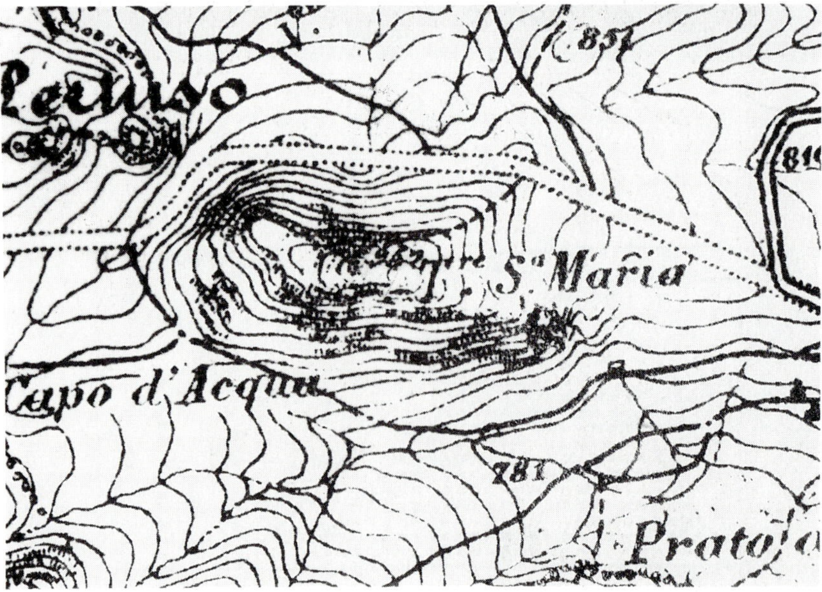

Fig. 6.18. Above: Aerial photograph of S. Maria dei Vignali wood taken in 1986. In this photograph the *tratturo* Lucera-Castel di Sangro is still visible (Authorization S.M.A. n.0/98). Below: Same area from a detail of a map from 1872 (scale 1:50000). The edge of the wood on the map was symbolized by small circles which excluded the area of ancient pasture of S. Maria that today is covered by oaks, maples and hornbeams. Ancient pathways are still visible. (*Dai tipi dell'Istituto Geografico Militare,* Autorizzazione n. 3562 del 21/02/1992.)

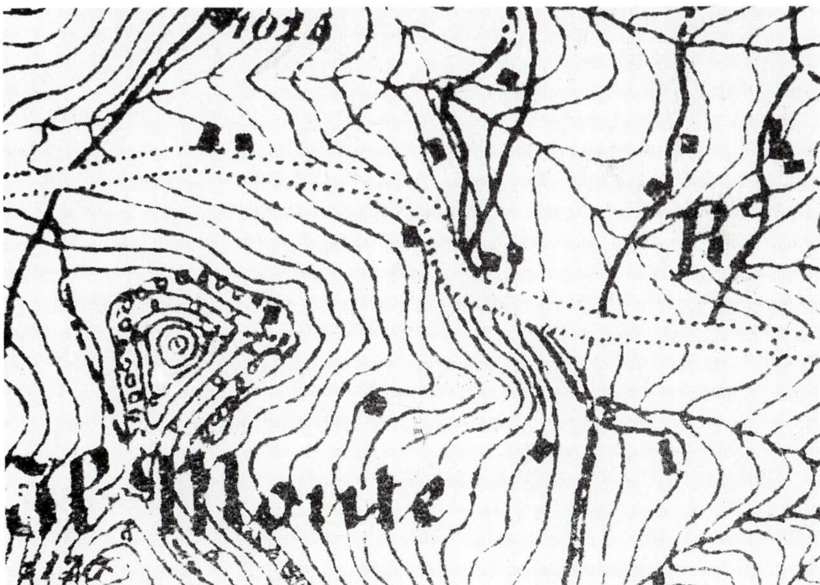

Fig. 6.19. Above: The *tratturo* Celano-Foggia and Il Monte wood in an aerial photograph of 1986. Scattered trees and shrubs are visible inside the area of the *tratturo* as well as fields. Below: Detail of the same area from the source cited in Fig. 6.18. Today the wood has regenerated naturally over much of the area between the old stand and the *tratturo*. (*Dai tipi dell'Instituto Geografico Militare*, Autorizzazione n. 3562 del 21/02/1992.)

are included to show how a comparison of old maps and aerial photographs is a useful method of evaluating forest landscape changes. These figures, all from the Apennines, show different rates and extent of woodland change.

Conclusions

The data suggest that livestock grazing and specific local practices used to modify stand structure and canopy density were the basis of forest management in Molise. Probably all woodland in the first half of the 19th century was used as wood pasture. Some pasture was rented out and some was subject to common rights. These management practices protected the forests from agricultural pressure and massive felling. This period was, in fact, characterized by a multiple land-use system in which coppice, grass and pollards were all important elements. There was a low level of wood cutting, which was sufficient to satisfy the essential needs of the local population. Legislative regulations that called for the uneconomic management of municipal woodland were frequently ignored.

The gradual decline of transhumance and the great social and economic changes that took place after the unification of Italy resulted in the widespread devastation of the woodland growing in the coastal and hilly part of Molise. At the same time, with the decline of agriculture in the mountainous western part of the region, there was a gradual spread of naturally regenerated woodland on once cleared arable fields and pasture. Most of this regeneration is by broadleaved tree species, which have a vigorous power of regeneration through coppice regrowth. The data provided in this study should lead to further research on the historical ecology of surviving old woodland and also of the new areas of woodland, whether afforested by natural or artificial means.

Archive Documents and Maps

Archivio di Stato di Campobasso 'Intendenza' Boschi, buste 192–926.

7

Pine Plantations on Ancient Grassland: Ecological Changes in the Mediterranean Mountains of Liguria, Italy, During the 19th and 20th Centuries

D. MORENO, G.F. CROCE, M.A. GUIDO AND
C. MONTANARI

SUMMARY

An area of ancient grassland growing on mountains facing the Ligurian Sea has been continuously grazed since at least 1500 AD under a common land-use system known locally as *communaglia*. The grassland stretches from the edge of permanent cultivated land up to the watershed. The system allowed free grazing and hay making; in addition there was some long fallow cultivation and permanent olive and chestnut groves. In the 1930s several areas were planted with *Pinus pinaster* and *Pinus nigra* by the State Forest Service. Changes in the ecology brought about by this afforestation are studied by comparing historical maps and photographs with surveys of the present flora and pollen analysis of the soil profile.

INTRODUCTION

In a previous research project, the use of historical photographic records as sources for the study of local vegetation history was evaluated (Moreno and Montanari, 1988). The combination of single dated photographs with fieldwork enables the measurement of the timescale of development and degeneration processes, the dating of biological and geographical features in the present landscape and the evaluation of phytosociological dynamics. In Italy, historical photographs of woodland and grassland were taken by the State Forestry Administration from the 1870s onwards. It was at that period that afforestation with pine became important, both in order to protect the soil and for economic reasons (Croce and Moreno, 1988). Historical photographs have been used in conjunction with historical documents and ethnographical, archaeological and botanical evidence.

In this chapter, in which we report on work in progress, the land-use history of the mountainous area surrounding Genoa (Fig. 7.1) before the pine plantations were made is considered. An important part of this type of research is the identification of historical and traditional land-management practices that were the basis of local land-use systems. These local systems have often been misunderstood, ignored or even despised, by outside contemporary commentators writing reports and compiling records. As a result the geographical, legal and agricultural terminology used for different types of land-use in 18th and 19th century records needs to be very carefully evaluated.

In addition, the ways in which recent, historical and ancient grassland sites can be detected, dated and plotted on the basis of very different types of archival and field evidence are shown. A preliminary attempt is made to gain an understanding of the modern ecology of the herbaceous layer of the grassland sites as a function of the history of local management. It is hoped that, in the future, fruitful contacts established with researchers in environmental archaeology will increase the usefulness of botanical evidence in the historical period.

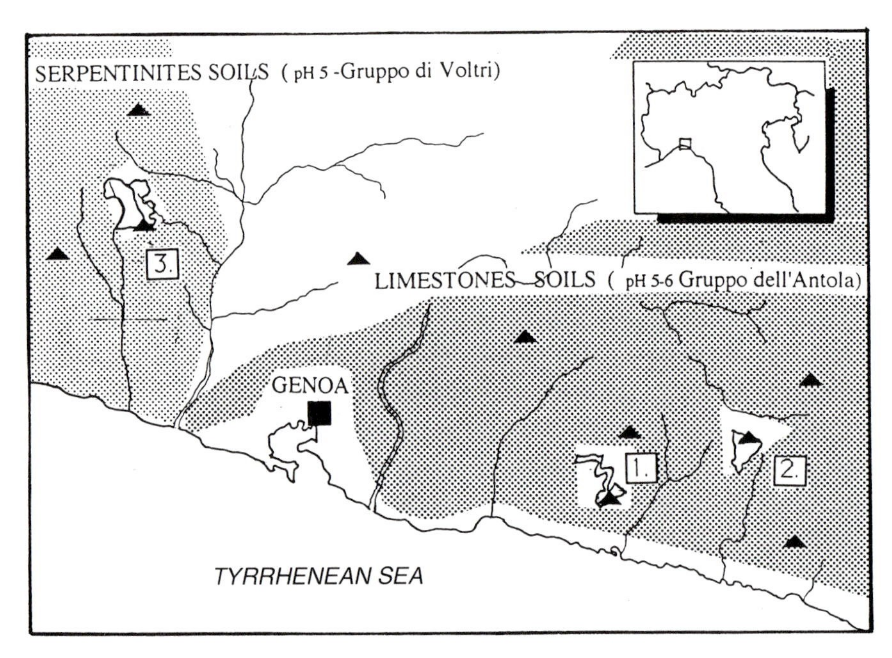

Fig. 7.1. Location of the three pine plantations studied. [1]: M. Fasce-M. Moro (ave. height 200–550 m). [2]: M. Cordona (ave. height 425–790 m). [3]: Ramasso (ave. height 200–650 m).

THE MODERN AND HISTORICAL ECOLOGY OF PLANTATION SITES

There have been two main phases of widespread planting by successive state forestry administrations in Liguria. The first was from Napoleonic times to 1830, the second was from 1870 up to the middle of the 20th century. In this section, three pine plantations that were all established after 1870 in the mountainous area surrounding Genoa are discussed. Figure 7.1 shows the location of the three plantations and sketches the two principal rock types and related soils. In this area, weathering has decalcified the limestone soils and their pH is slightly acid.

Gentile (1984) has identified two phytoclimatic vegetation belts for the south-facing slopes of this group of mountains. The littoral (Mediterranean) belt is found between 0 m and 300–400 m above sea level; the sub-Mediterranean belt is found between 400 m and 800 m above sea level (Fig. 7.2). The vegetation is generally thermophilous, although species associated with mesophilous woodland (*Fagetalia sylvaticae*) and grasslands (*Molinio-Arrhenatheretea*) can also be found (Nowack, 1987). *Pinus pinaster* is now an important element of the vegetation in both zones.

The current species composition of the grassland is shown in Table 7.1. Data were obtained from 34 phytosociological surveys in the area, but only the most frequent species are noted. The floristic and synecological features of the vegetation enable us to classify it as *Mesobromion* grassland. Environmental and

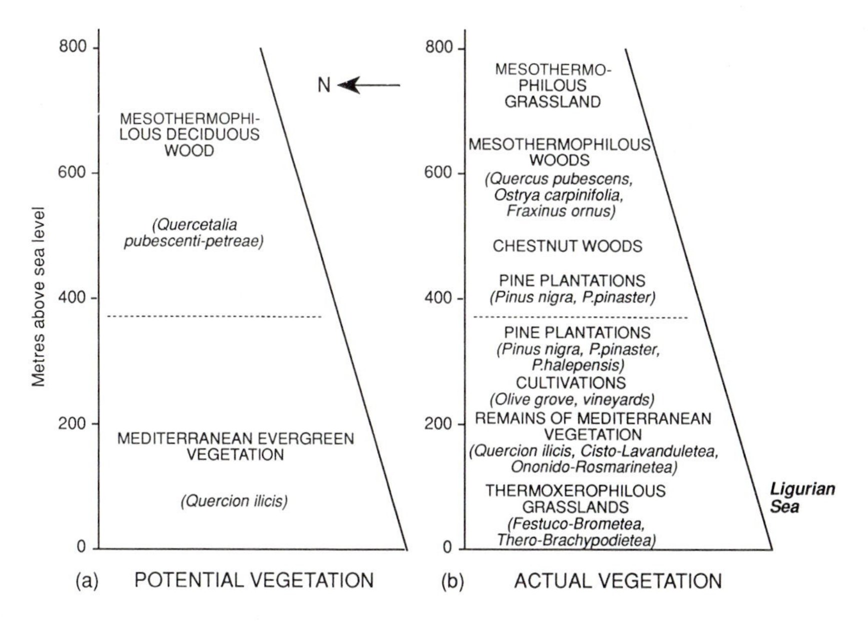

Fig. 7.2. Altitudinal zonation of the potential (a) and actual (b) vegetation of the areas studied.

Table 7.1. The species composition of modern grassland in the Fasce–Moro and Ramasso mountains.

Species	Percentage frequency[a]	Cover value[b] range
Brachypodium pinnatum s.l.	100	1 – 5
Bromus erectus Hudson	88.23	+ – 4
Dactylis glomerata L.	79.41	+ – 3
Helianthemum nummularium Miller ssp. *obscurum* (Celak.) Holub	79.41	+ – 2
Peucedanum cervaria L.	67.64	+ – 3
Lotus corniculatus L.	67.64	+ – 1
Carex flacca Schreber	64.70	+ – 1
Trisetum flavescens (L.) Beauv.	64.70	+ – 2
Leucanthemum vulgare Lam.	61.76	+ – 2
Lathyrus latifolius L.	55.88	+ – 2
Anthoxanthum odoratum L.	55.88	+ – 3
Galium corrudifolium Vill.	52.94	+ – 2
Sanguisorba minor Scop.	50	+ – 1
Polygala nicaeensis Risso	50	+ – 1
Chrysopogon gryllus (L.) Trin.	50	+ – 3
Teucrium chamaedrys L.	50	+ – 1
Briza media L.	47.05	+ – 1
Onobrychis viciifolia Scop.	47.05	+ – 1
Plantago lanceolata L.	47.05	+ – 1
Agrostis tenuis Sibth.	47.05	+ – 2
Genista tinctoria L.	44.11	+ – 1
Festuca rubra L.	44.11	+ – 3
Hippocrepis comosa L.	44.11	+ – 1

[a]These presence figures have been calculated on the basis of 34 phytosociological surveys randomly located in the studied area.
[b]Minimum and maximum cover value (abundance/dominance) is indicated, according to Braun-Blanquet (1964): + = covering less than 1%; 1 = covering from 1–5%; 2 = covering at least 25%; 3 = covering from 26–50%; 4 = covering from 51–75%; 5 = covering from 75–100%.

historical factors have produced a species-rich composition and the most mesophilous species of thermoxerophilous grassland (*Festuco-Brometea*) are often found growing with the more thermophilous species characteristic of fertilized meadows (*Molinio-Arrhenatheretea*). This mixture of plants that is characteristic of the Mediterranean and central European flora in places so close to sea level is peculiar to these mountains of coastal Liguria.

The general structural changes following the establishment of pine plantations were examined by comparing the range of species found in pine plantations and neighbouring grasslands. In addition, photographic evidence was used for plantations made in the 1930s (Moreno *et al.*, 1992, Figs 1A–1D, pp.162–163). Structural modifications brought about by the establishment of a tree canopy are shown in Figure 7.3. There was less of a decrease of Hemicryptophyta than expected and Therophyta, although halved, are still present. Chamaephyta have not changed, but there has been a remarkable increase in Geophyta and Phanerophyta. Pines have not been taken into account.

The general pattern of changes to the phytosociological assemblage that took place following the growth of the pines is shown in Figure 7.4. As expected (Biondi and Ballelli, 1973), in the pine plantations there is a marked decrease in species typical of thermophilous and mesophilous grassland (*Thero-Brachypodietea, Festuco-Brometea, Molinio-Arrhenatheretea*), while the decrease in Mediterranean low-growing shrubs (*Ononido-Rosmarinetea, Cisto-Lavanduletea*) and species that are characteristic of woodland borders (*Trifolio-Geranietea*) is only slight. On the other hand, species typical of evergreen Mediterranean vegetation (*Quercetea ilicis*) and of mesothermophilous sub-Mediterranean and continental woods (*Querco-Fagetea*) show a strong increase.

Figures 7.5 and 7.6 provide a more detailed picture of the present-day ecology of individual sites. Figure 7.5 shows the differences in the number of species found in modern pine plantations, heathland and grassland, while 7.6 shows the relationship between the density of pine cover and the number of plant species. As might be expected, there is an inverse correlation between the density of the arboreal layer and the number of plant species. However, two sites stand

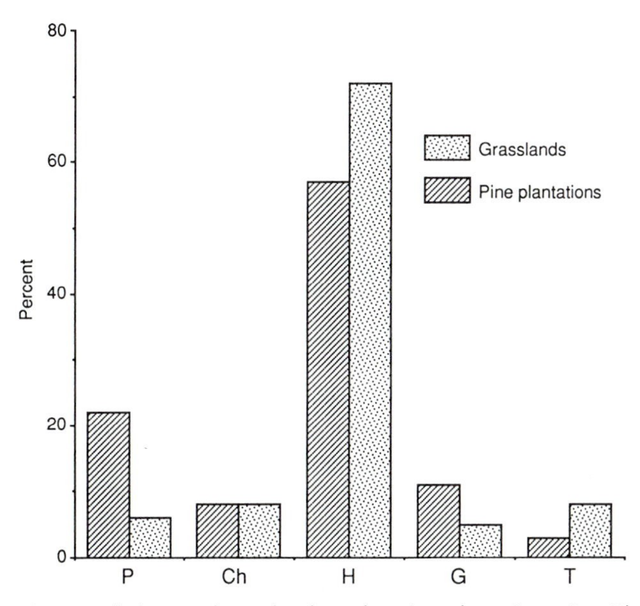

Fig. 7.3. Structural changes brought about by pine plantations. P = Phanerophyta; Ch = Chamaephyta; H = Hemicryptophyta; G = Geophyta; T = Therophyta.

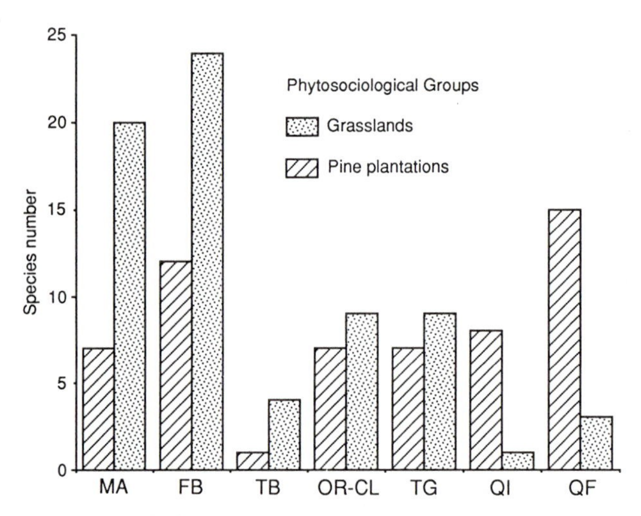

Fig. 7.4. Comparison of phytosociological groups in pine plantations and nearby grasslands. MA = *Molinio-Arrhenatheretea;* FB = *Festuco-Brometea;* TB = *Thero-Brachypodietea;* OR = *Ononido-Rosmarinetea;* CL = *Cisto-Lavanduletea;* TG = *Trifolio-Geranietea;* QI = *Quercetea ilicis;* QF = *Querco-Fagetea.*

out as exceptional. Site 1 (Fig. 7.6) is a *Pinus pinaster* plantation in the Ramasso area that was established 110 years ago on ancient common grassland on a steep, possibly eroded, slope with a serpentinite soil. Today it has an herbaceous layer dominated by *Sesleria appenina* with few accompanying species. By way of contrast, Site 8, a pine plantation established in 1935 on former meadowland in the Fasce-Moro Mountains retains a high number of species.

Grassland Archaeology and Documentary Evidence: Plantations in the Fasce and Cordona Mountains

It is generally agreed by biogeographers that grasslands in the study area are of secondary origin. This hypothesis was already clearly established in the local agricultural literature of the 19th century. It is now possible, using archaeological techniques, to provide detailed information about the former woodland in this area. There has recently been rescue archaeology work on the mountain route of a methane pipeline passing through historic grassland on the edge of the Mount Cordona pine plantation. The site is 760 m above sea level and has a north-west aspect. The excavation provided a stratigraphically documented buried profile of the woodland 'climax' palaeosoil. The fabric of the natural argillic forest soil has been examined through micromorphological analysis.

Under a charcoal-rich layer (UN2) (radiocarbon dated 2775 +/-150 years BP) the remains of a woodland 'climax' brown soil appears to have been

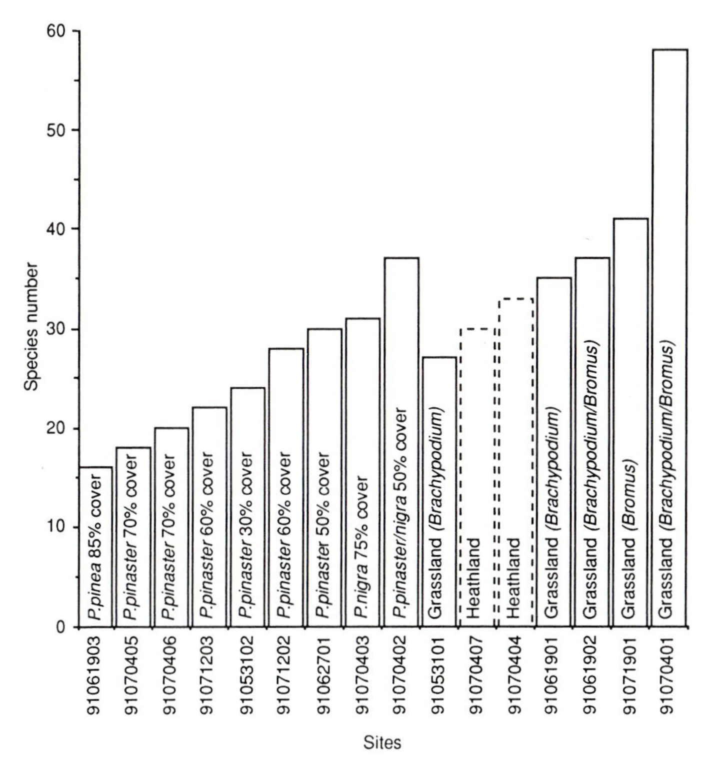

Fig. 7.5. The number of species in modern pine plantations, heathland and grassland.

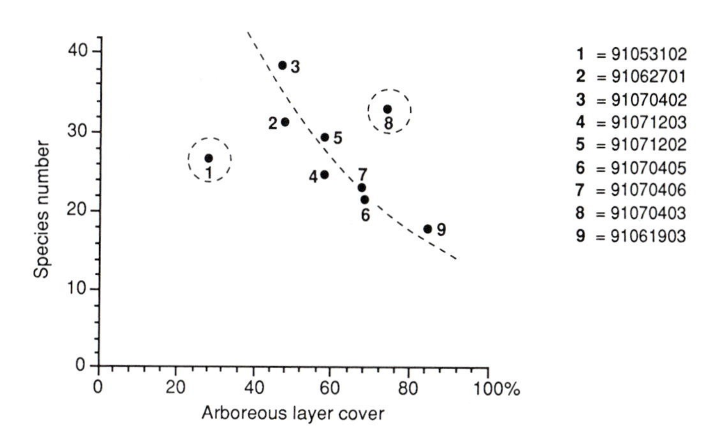

Fig. 7.6. The relationship between the number of species and the percentage canopy cover.

truncated in the Late Bronze Age by the grubbing up of trees for agricultural or pastoral purposes with the aid of fire (Cremaschi *et al.*, Fig. 3.5, pp.44–45). It appears that there is a secondary forest soil in the upper layers. Pollen from these poor upper deposits (undated) are mainly Gramineae but the presence of *Fraxinus*, *Alnus*, *Castanea*, *Corylus*, *Pinus*, *Quercus*, *Erica* and *Calluna* may suggest the development of tree and shrub cover. Unfortunately, however, it is impossible to specify whether this was of local or regional extent (Moreno *et al.*, 1992, p.1). At the nearby site of Mount Borgo (728 m above sea level), which has a similar modern ecology and topography, a climax argillic forest soil has been documented by micromorphological analysis. The related major deforestation and cultivation phase has been dated using archaeobotanical and archaeological findings to the Copper and Early Bronze Age (Macphail, 1990).

It is hoped that the study of the land-management history of individual sites will, in the future, allow us to say whether the area remained an undisturbed grassland following the clearance of woodland. At present, we can only state that the grasslands of the Fasce-Cordona Mountains were continuously used as winter pasture for sheep and goats from at least the 15th to the early 20th centuries as part of the local commonland system known as *communaglia* (Moreno and Raggio, 1991; Moreno *et al.*, 1992).

Archaeological artifacts from deserted medieval villages can also provide evidence for the land-management history of the eastern Ligurian Apennines. Indeed, a theory that the production of hay became important only very late because of the rarity of iron and the technical difficulty of producing scythe blades in the medieval period has recently been disproved by the discovery of sickles and scythe blades found at the deserted village of Zignago, which were dated to the first half of the 14th century (ISCUM, 1987, Figures 33–36).

Although the shepherd economy collapsed during the second half of the 19th century, 5000–6000 sheep and goats were still using the winter grazing of the M. Fasce-Cordona *communaglia* at the beginning of the present century, and the last flocks of local Genoese sheep breeds could still be found on this mountain in 1950. Transhumance roads, pastoral corbelled huts and drystone dykes enclosing olive and sweet chestnut groves, meadows and horticultural parcels of land remain as standing archaeological features of the grassland landscape. These parcels were individually owned properties within the common land (known as *usurpazioni* in 17th- and 18th-century records).

The areas both to the west and east of Genoa were part of the *communaglia* system during the post-medieval period, and acted as outfield areas for the intensive horticulture (with vineyards, orchards and olive groves) of the peri-urban Genoese area: the *agricoltura di villa*. In the mountain commons, the boundaries between meadows, wooded meadows and pasture with trees or shrubs were indefinite and movable during the post-medieval period. The *communaglia* system implies the multiple use (agrosylvo-pastoral) of the vegetation.

Hay was a very marketable commodity and this encouraged the improvement or *domesticazione* of common grassland. Long fallow fields were sown with Gramineae such as *Secale*, *Avena* or *Hordeum*, or Leguminosae such as *Ervum*. The existing vegetation in these parcels was stripped off and then dried and burned, a practice called in the local post-medieval records *ronco* or *fornelli* (Moreno, 1990).

Unfortunately, tithe surveys (*mappe catastali*) of the 18th and 19th centuries do not exist for the Apennine commons because such land was not subject to state fiscal control. It is possible to produce large-scale maps of the historical distribution of broad vegetation types. Such a map has recently been published showing the distribution of meadows (*prati*), pastures, heathland and wooded meadows (**gerbidi**) according to the classification and scale of the printed edition (1852) of the *Gran Carta degli Stati Sardi in Terraferma* (Moreno *et al.*, 1992, Fig. 64). However, this kind of archival source is quoted here as an example of a type of source that is too generalized to be of much value for the purposes of historical ecology (Rackham, 1986).

In the mountainous area under discussion, common rights of making hay, collecting herbs and grazing were still held by certain parishes as late as the second half of the 19th century despite the establishment of communal boundaries after 1815. Figure 7.7 shows the collapse of the common grassland management system and the subsequent land allotment in the Commune of Quarto between 1798 and 1938. In this period, animal production became centred on stalled cattle fed on herbs cut throughout the year in addition to seasonal hay production. Over the past 50 years, most trees – mainly groves of chestnut and mixed oakwood – have been grubbed out and the area of grassland has been extended dramatically (Fig. 7.8). Hay making is now carried out on individual farms, although it is interesting to note that it is still mown in individual strips known as *pezze* as in the former common grassland system.

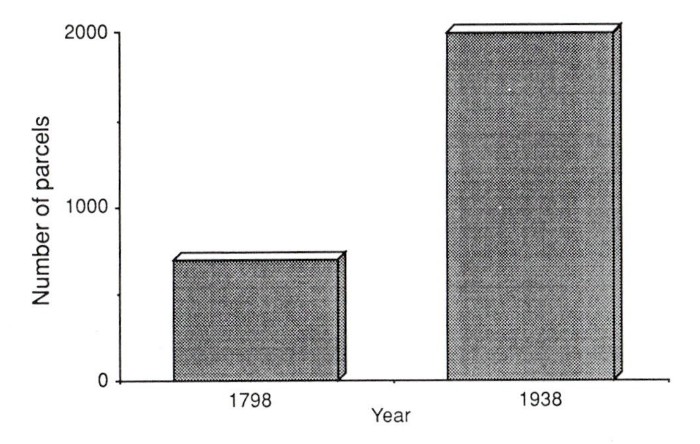

Fig. 7.7. The increase in the number of enclosed land parcels and the decline of commonland 1798–1938.

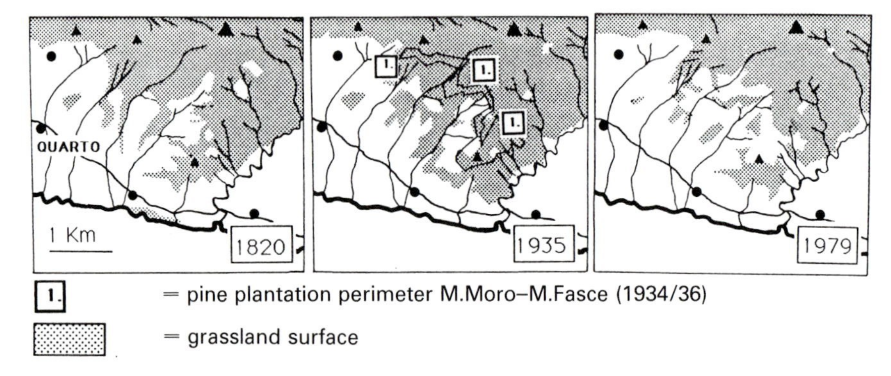

Fig. 7.8. The distribution of grassland on the southern slopes of the Fasce-Moro Mountains in 1820, 1935 and 1979. The maps used as sources for this figure and Table 7.3 are as follows. **1820**: manuscripts at a scale of 1:9450 'Tavolette di campagna' by the Corpo di Stato Maggiore of the Kingdom of Sardinia now in the Archivio Istituto Geografico Militare (IGM), Firenze, Cart. 31/251/74-75-72 (based on field survey); **1852**: Gran Carta degli Stati Sardi in Terraferma, 1:50000 scale, Corpo Reale di Stato Maggiore, Torino (Fogli 67-68-76) (field survey); **1878**: First edition of 1:25000 scale IGM tavolette NERVI-RECCO-S. OLCESE (field survey); **1935**: Second edition of 1:25000 scale IGM, tavolette NERVI-RECCO (aerial photographic survey); 1979: 1:10000 scale, Carta Tecnica Regionale, Liguria, elemento 231010-214130 (aerial photographic survey).

Detailed information, collected for fiscal purposes, is available about the state of the herbaceous layer since the enactment of the *Nuovo Catasto* in 1935. Using the Commune of Quarto as an example, it can be seen how the different values and classes of the herbaceous layer were identified in accordance with the traditional hand tools (of medieval origin) used and the different dates on which the hay was cut (Table 7.2; data from Ascari, 1938). The maps in Figure 7.8 show the changing distribution of grassland on the southern slopes of the Fasce-Moro Mountains between 1820 and 1979 based on historical maps. The existence of a set of manuscript large-scale maps (1:9450) surveyed in 1820 by the *Corpo Reale di Stato Maggiore* (Staff Officers of the Army of the Kingdom of Sardinia) has allowed the correction of the printed edition of 1852. These data have been used in conjunction with land-use class information on modern maps produced at almost the same scale (1:10000) by the Regional Government of Liguria from aerial surveys (Carta Tecnica Regionale, CTR, 1979).

The same sources allow us to distinguish between recent and ancient pine woodland. Table 7.3 shows land-use history for the 46 sites found bearing pine trees in the 1979 Regional Aerial Photography Survey (CTR, 1979). Preplantation sites are classed as ancient if trees or woodland were shown on the 1820 manuscript maps. These areas could be classed as ancient semi-natural woodland (Watkins, 1990, p.12) but we consider that it would be better if these Mediter-

Table 7.2. The classification of the herbaceous layer according to the *Nuovo Catasto* and hay production practices in the Commune Quarto (*c.* 1935).

Classification	Turf condition	Exposed rock/debris	Cut period	Tools employed
Meadows (1st class)	Continuous	0–15%	May/June	Scythe (dial. skuriata)
Meadows (2nd class)	Continuous	15–20%	July	Scythe & sickle (dial. mesuya)
Meadows (3rd class)	Shrubs	> 30%	July	Sickle
Wooded/meadows	250 trees/ha		May and October	Scythe
Olive groves	300–500 trees/ha	Terraced	May and October	Sickle
Chestnut groves	300–400 trees/ha	Terraced	May and October	Sickle
High forest	1000–1200 trees/ha		May and October	Sickle

Source: M.C. Ascari, 1938.

ranean woodlands with a known post-medieval management history were termed historical.

RECENT VEGETATION CHANGES: THE BOSCO RAMASSO (1812–1991)

In this section, vegetation change along a transect in the Ramasso Valley (Fig. 7.9, Transect B; Moreno and Croce, 1993) is assessed. The main sources used were the *plan parcellaire* of the 1812 cadastral map, surveyed during the Napoleonic administration and the *Consegna dei Boschi* enacted by the Kingdom of Sardinia Forestry Administration in 1820. In addition, use was made of a very detailed map (1:8000) published in a parish history in 1873 and originally drawn up as evidence for a legal dispute with commoners from neighbouring parishes. The transect is shown in Figure 7.10.

This research confirmed that care has to be taken to compare field evidence with the documentary descriptions. The common land shown as gerbidi on the 1852 topographical map, for example, is depicted as grassland in the 1820 survey and plotted as *beni coltivati a foraggi* on the 1873 map. Other more detailed mid-19th century records indicate that the vegetation concerned was a special type of grassland with trees. Arboreal species such as *Ostrya, Fraxinus* and *Alnus* were cut back to the ground every year when the hay was cut, thus providing valuable additional dry fodder (see Chapter 5). Functionally, these

Table 7.3. List of ancient and recent pine woodland sites on the southern slopes of the Fasce-Cordona mountains. P = pine; P* = planted pine; C = chestnut; T = tree cover; G = grassland; O = olive; M = meadow; W = woodland. Date: r = recent, a = ancient, u = uncertain.

Site	Height average (m)	Height class	Slope	Land-use history				Date
				1979	1878	1852	1820	
1	200.0	2	NE	PC	C	W	T	a
2	270.0	2	E	P	T			a
3	500.0	4	W	PC	T	G	G	u
4	300.0	3	NE	PC	G	G	G	r
5	275.0	2	SE	PC	T	G	P	a
6	350.0	3	NE	P	T	G	G	r
7	412.5	3	S	P	GT	G	G	r
8	325.0	3	W	PC	GT		G	r
9	375.0	3	W	PC	GT		G	r
10	300.0	2	W	P	G		T	a
11	387.5	3	W	PC	G		G	r
12	385.0	3	W	PC	G		G	r
13	225.0	2	E	P	GT	G		u
14	225.0	2	NE	PC	C	W	TG	a
15	150.0	2	SE	PC	O	O		u
16	87.5	1	SE	P	GT	G		u
17	125.0	1	W	PC	GT	C	GT	a
18	187.5	2	W	PC	GT	G	GT	a
19	187.5	2	W	PC	T	O	T	a
20	362.5	3	W	PC	GT		G	r
21	312.5	2	W	PC	T		G	r

22	100.0	1	SE	P	G		G	r
23	125.0	1	S	P	G		G	r
24	262.5	2	W	PC	GT	O	G	r
25	200.0	2	W	PC	GT		G	r
26	50.0	1	S	PC*	T	O	T	a
27	462.5	4	NW	PC	C	M	T	a
28	387.5	3	NW	PC	C	W	T	a
29	412.5	3	NW	PC	C	W	T	a
30	350.0	3	NW	PC	C	W	T	a
31	250.0	2	NW	PC	C	W	T	a
32	162.5	2	N	PC	C	W	T	a
33	262.5	2	S	P*	G		G	r
34	365.0	2	S	P*	G		G	r
35	375.0	3	W	P*	G		G	r
36	400.0	3	W	P*	G		G	r
37	470.0	4	W	P*	G		G	r
38	187.5	2	E	P	T	G	P	a
39	87.5	1	SE	P	T		P	a
40	25.0	1	S	PC*	T	O		u
41	587.5	4	S	P*	G	G	G	r
42	665.0	5	S	P*	G	G	G	r
43	675.0	5	S	P*	G		G	r
44	750.0	5	SE	P*	G		G	r
45	750.0	5	E	P*	G	G	G	r
46	400.0	3	E	P*	G		T	r

Sources: See caption to Fig. 7.8.

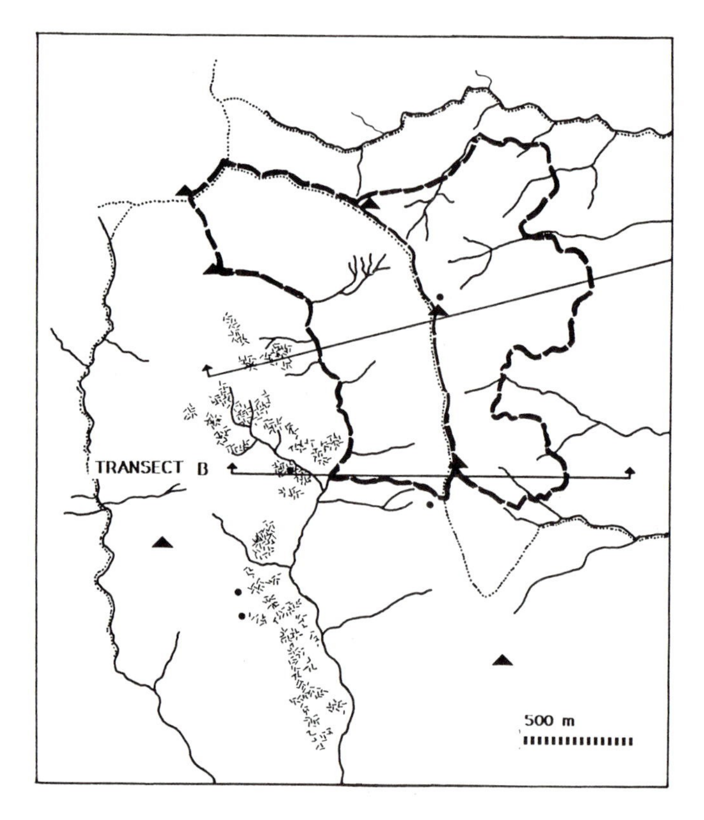

Fig. 7.9. Map of the Ramasso Plantation. ——— Plantation perimeter (recent pine cover) (1880–9120). .... Boundary of the 'commune' (1859). Transect A and B locations. • Phytosociological survey sites. Area of the historic *P. pinaster* wood ('ceduo da fornace' 1820) (historical pine cover).

areas were grassland but legally they were classed as woodland. In the Eastern Riviera this practice was still being carried out on privately owned grassland as late as 1960. Under this system of land use, the date of hay cutting was determined by the species of tree. The leaves of *Ostrya*, for example, are only palatable for cattle before the end of July. The cutting was performed with sickles. In the Ramasso common grassland, seasonal uprooting of shrubby species (such as *Erica*, *Calluna*, and *Spartium*) was a regulated practice until the 1920s. These practices produce a grassland-meadow aspect to the landscape that provides a seasonally and cyclically varying vegetation. This kind of grassland disappeared in the Ramasso perimeter under pine and broadleaved (*Castanea, Robinia*) plantations.

On the western side of the Ramasso Valley (see Fig. 7.9) there are some areas of *Pinus pinaster* woodland that documentary evidence indicates have been in existence since at least 1812–1820. This woodland is described on Transect B as

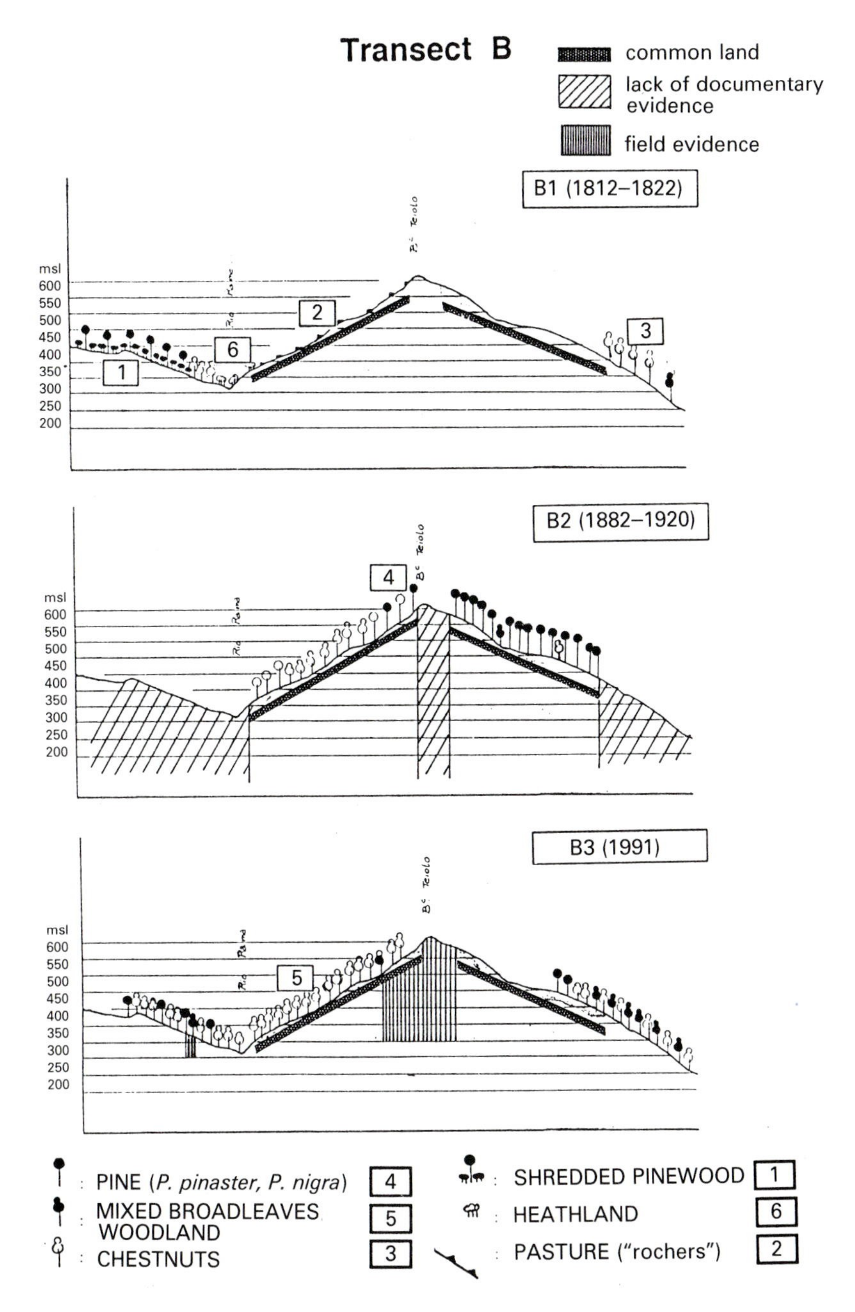

Fig. 7.10. Transect B in the Ramasso Plantation.

ceduo da fornace con pini (or in contemporary records as *pini cedui nei rami*). It was a type of privately owned shredded-pine wood in which the pines were felled and replanted on a 20–25-year cycle. Single trees, or clusters of trees, were planted between the existing shredded trees. There was also a dense, shrubby layer of arboreal species coppiced on a very short cycle of 3–5 years (Fig. 7.11). The pine woodland on these sites has been in existence for at least 200 years. Moreover, if the management practice can be related, as the local name suggests, to the period when lime pits were active in the area, then it could be 450 years old. Since the abandonment of shredding practices in the 1940s, the *ceduo da fornace* on these slopes is gradually changing into broadleaved woodland of chestnut in the north, and Turkey oak in the south, with a sparse scatter of pine trees.

The survival for so long of these pine woodlands appears to argue against the received biogeographical wisdom that pine is expected to be replaced following natural secondary succession by broadleaved woodland. However, it must be remembered that pine in Liguria was commonly regarded in peasant

Fig. 7.11. Shredded pinewood sketched from Giacobbe, 1942.

culture as a species that invaded cultivated groves of sweet chestnut, figs and olives. This belief was recorded, and shared, by authors such as Giacobbe (1942) who reported a local vernacular proverb that 'pine is the weed of woodland'.

Different sources, including archive records and field observations, show the demise of many local land-management practices affecting commonland vegetation (Fig. 7.12). Such intensive land-management practices meant that the

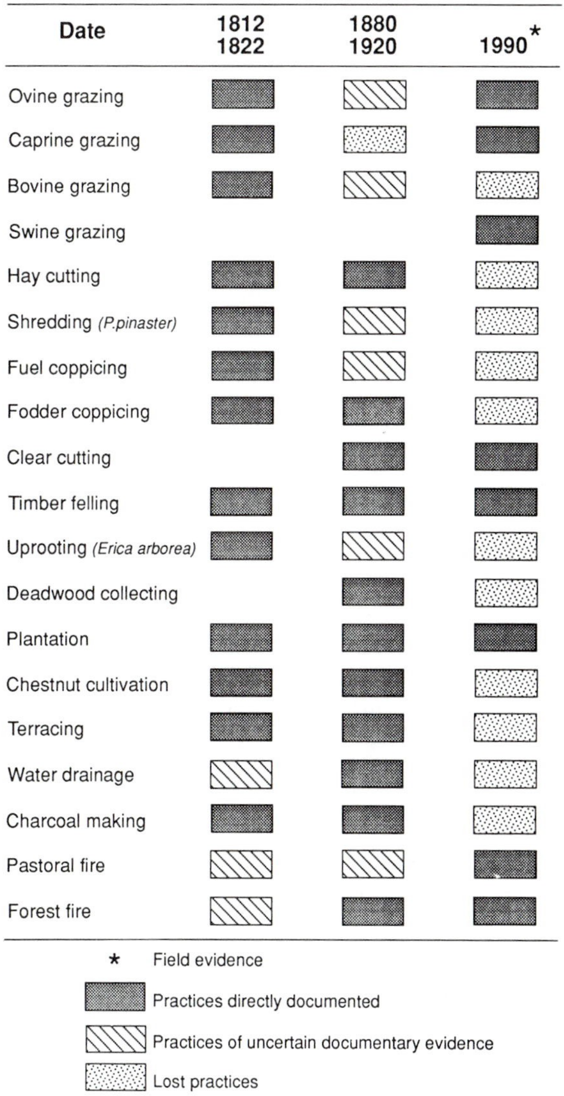

Date	1812 1822	1880 1920	1990*
Ovine grazing	documented	uncertain	documented
Caprine grazing	documented	lost	documented
Bovine grazing	documented	uncertain	lost
Swine grazing			documented
Hay cutting	documented	documented	lost
Shredding *(P.pinaster)*	documented	uncertain	lost
Fuel coppicing	documented	uncertain	lost
Fodder coppicing	documented	documented	lost
Clear cutting		documented	documented
Timber felling	documented	documented	documented
Uprooting *(Erica arborea)*	documented	uncertain	lost
Deadwood collecting		documented	lost
Plantation	documented	documented	documented
Chestnut cultivation	documented	documented	lost
Terracing	documented	documented	lost
Water drainage	uncertain	documented	lost
Charcoal making	documented	documented	lost
Pastoral fire	uncertain	uncertain	documented
Forest fire	uncertain	documented	documented

* Field evidence

Practices directly documented

Practices of uncertain documentary evidence

Lost practices

Fig. 7.12. Local land-management practices affecting the Ramasso commonland vegetation.

area supported many people under the *communaglia* multiple-use system before the first half of the 19th century and the later appearance of plantation forestry. No central or local archival records exist for the area after 1950 and, as a result, the recent history of grassland and woodland management can only be reconstructed from field and oral evidence. Some practices, such as swine grazing, which was introduced in the 1960s, are of recent origin. Uncontrolled forest fire is now assumed to be the principal environmental factor affecting all abandoned land and pine woodland in the region.

Conclusions

These conclusions are both preliminary and speculative. In this chapter the terms ancient and historical have been used interchangeably as have the terms recent and secondary. Multiple land-use systems similar to those discussed in this chapter have been applied to three-quarters of the land area of the north-western Apennines since the first half of the 19th century (Ullmann, 1967). It is therefore important to discover how the traditional woodmanship practices of the earlier period changed and developed into those current in the first half of the 20th century. Early in the present century historical practices were still carried out, but their economic and ecological significance was very different to that of similar practices in the early 19th century. The traditional way of managing pine trees, for example, has changed substantially over the past 200 years. Moreover, there is local variation, with the practice of shredding and replanting pine trees being found in the western, but not the eastern mountain ranges. The ecological behaviour of the pine also varies. At some sites, the pine replaces broadleaved trees such as chestnut, while at other sites the pine is replaced by broadleaved trees. Preliminary observations of the modern sites bearing pine trees (see Table 7.3) shows that by 1979 pine had invaded all the sites listed in 1878 as sweet chestnut groves. This is in accordance with an old peasant proverb, but appears to contradict conventional ecological hypotheses.

Acknowledgements

This chapter is dedicated to Professor Rodolfo E.G. Pichi Sermolli, a field botanist who nurtured the authors' own approaches to the world of plants, on his 80th birthday. We would like to acknowledge the help of O. Raggio in providing information and photographs and the assistance of P. Rowley-Conwy, J.G. Lewthwaite and D. Coombes in editing the text.

8

Ecological Effects of a Less-intensively Managed Afforestation Scheme on Dartmoor, South-west England

S.J. ESSEX AND A.G. WILLIAMS

SUMMARY

This paper reviews some of the ecological effects of a less-intensively managed conifer plantation around Burrator Reservoir, Dartmoor, between 1916 and 1991. Well-managed afforestation schemes are known to cause profound changes to the ecosystem but the ecological effects of a lower intensity of management are not well documented. The human management decisions concerning the development of the afforestation are described together with their resultant ecological changes. This research focuses particularly on the period 1976–1991 and shows that the less-intensive management of the forests at Burrator resulted in a woodland ecosystem with a relatively diverse flora. The influence of afforestation on the soils and stream chemistry of the area was found to be relatively limited. Few observable changes have occurred recently following clear-cutting and replanting because of lags in the system. An assessment of the potential changes to forest management following water privatization is made, together with a consideration of the possible future ecological impacts.

INTRODUCTION

Intensive coniferous afforestation has been shown to lead to profound changes to both the living and the non-living components of an ecosystem (Nature Conservancy Council, 1986). Goldsmith and Wood (1983) particularly criticized the commercial plantations that dominate much of the British uplands on the grounds that they are only composed of one species of conifer and tend to be of a single age, thereby creating monocultures with a simple ecological structure. Existing flora suffers greatly from intensive afforestation. Mature stands are so

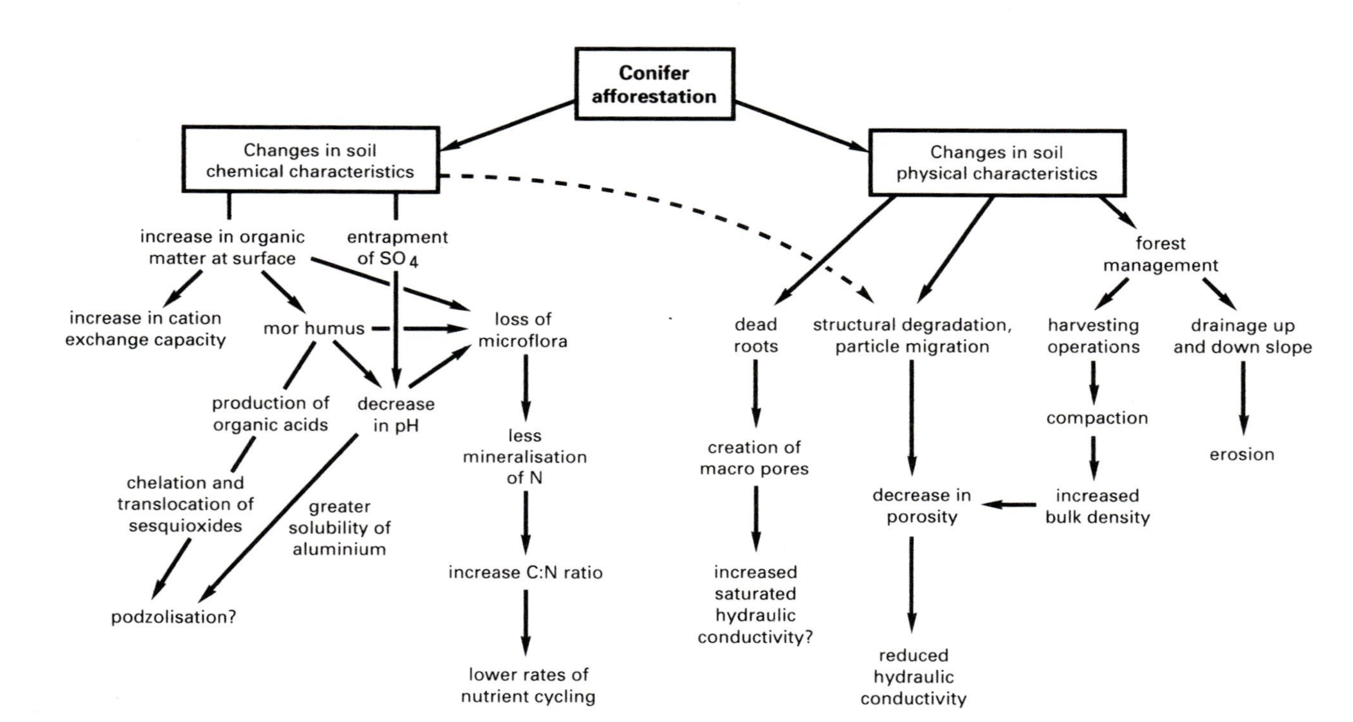

Fig. 8.1. Possible effects of conifer growth on soils.

dense that light levels at the ground surface are reduced to a minimum and hence the undergrowth is shaded out. Furthermore, during harvesting and subsequent planting much disturbance is caused to the ecosystem; this imposes sudden structural changes on any woodland flora and fauna that may have developed (see Chapter 2).

Conifer trees may also considerably affect the physical and chemical properties of soils, as shown in Figure 8.1. Nutrient cycling is limited, for example, because the acid conditions created by decomposing leaf litter limit the rate of nitrogen mineralization. Particular damage is caused during site preparation when less-weathered subsoil is brought to the surface. The drainage ditches often scour badly and transport water with high aluminium concentrations rapidly to streams. These high levels of aluminium in streams have been linked to reductions in macro-invertebrate populations and fish kills (Stoner and Gee, 1985).

In less-intensively managed regimes the ecological effects of coniferous afforestation can be expected to be more limited. Afforestation is generally on a smaller scale so that, within a given area, different blocks are at different stages of maturity. In the recently planted blocks or the post-mature areas wild vegetation is more common. Anderson (1979), for example, reported that long rotations of non-native coniferous species of over 80 years have been found to develop typical woodland habitats, that there were usually about 20 vascular plant species present and that the ground flora composition was similar to that found beneath oak.

The management of plantations in water-catchment areas has often suffered because the water authorities' primary objective was to maintain and protect water quality (Newton and Rivers, 1982; Essex, 1990). The management of such schemes has been subject to disruptions to the natural flow of activities, such as labour availability, market demands, logistical considerations and amenity pressures, which have had important implications on the overall profitability and function of the plantations. Afforestation can, therefore, be considered as less-intensively managed compared with commercial forestry. These circumstances have particular implications for parts of Britain, particularly in the uplands, since water companies own or manage about 9984 ha of forests, making them the fourth largest landowner of woodland in the UK (Harrison *et al.*, 1977).

This chapter describes the development and management of the plantation around Burrator Reservoir on Dartmoor, Devon. The aim of this research is to examine the ecological impact of an afforestation scheme where the intensity of management has been rather low and to explore the hypothesis that the effects of such management are strictly limited. The chapter attempts to make a direct association between human management decisions and the ecological consequences, thus integrating both human and physical geographical approaches to this particular issue.

CASE STUDY: BURRATOR RESERVOIR

Historical perspective

The Burrator Forests cover about 350 ha and are associated with a water supply reservoir serving Plymouth and South Devon. The reservoir is located about 20 km north of Plymouth and lies within the boundary of the Dartmoor National Park (Fig. 8.2). The construction of Burrator Reservoir in the Meavy Valley on Dartmoor in the late 19th century followed a protracted debate over the supply of water to Plymouth. Between 1825 and 1890, there were six proposals for a reservoir site on Dartmoor, which were all abandoned for various reasons. Eventually, in 1892, a site at Burrator Gorge was selected by Plymouth Corporation for a storage reservoir, which was inaugurated on 21st September 1898.

It was common practice for reservoired catchments to be afforested in the late 19th and early 20th centuries because of various perceived benefits for water management. Trees represented an ideal barrier for restricting potentially polluting agricultural uses and public access to catchment areas; prevented the reservoirs from silting up by holding the banks together; had perceived benefits for water supply in terms of both quantity and quality; and created employment and represented a source of regular income for the water managers from the sale of timber (Essex, 1990).

In line with this 'planting spirit', Plymouth Corporation had purchased all of the surrounding catchment area of Burrator (2200 ha) by 1916. However, afforestation of the immediate catchment area did not occur at once as the Corporation allowed the farmers in the valley to remain until 1925. Much of the immediate catchment area upon which the afforestation occurred was agricultural land that had been improved during the 19th century by the clearance of boulders, liming and ploughing. It is significant that the limed areas had a higher base status than the unimproved areas and ploughing may have broken up the indurated soil horizon to facilitate deeper rooting. These circumstances might have significant implications for the subsequent ecological effects of the afforestation, in that the soil conditions at afforestation were improved and became more resilient to change.

Afforestation of the catchment area

Figure 8.3 shows the spatial development of the afforestation from the initial planting around the reservoir itself in the 1920s, to the major period of planting in the valleys of the feeder streams during the 1930s and, finally, to minor infilling after 1960. By the late 1980s, the afforestation totalled 356.1 ha, which represented about 16% of the catchment area. The general composition of the plantations is shown in Figs 8.4 and 8.5. Coniferous trees formed most of the plantings (298.7 ha or 84%), reflecting the constraints of the soil conditions and

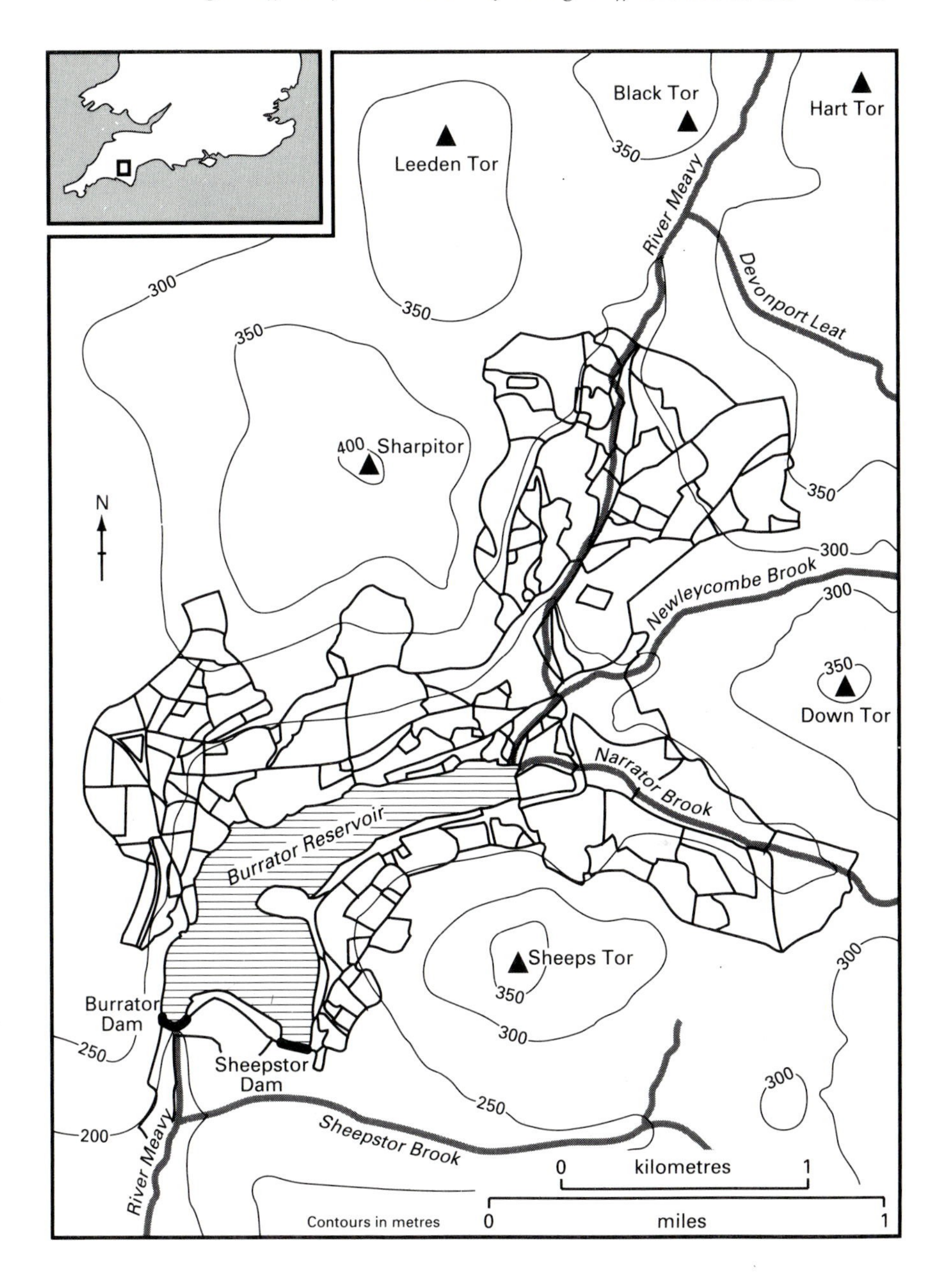

Fig. 8.2. Location of Burrator Reservoir and compartment outlines of the afforestation.

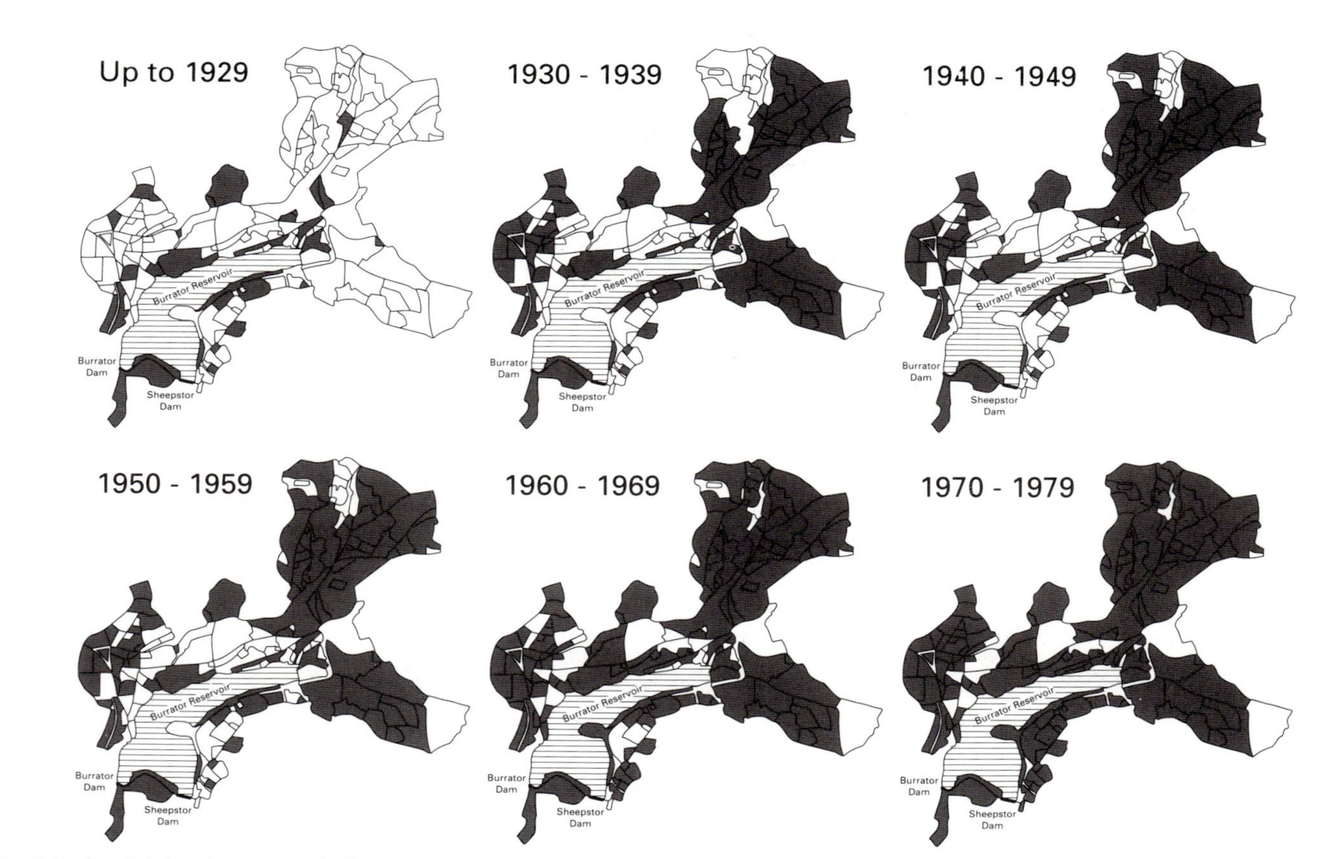

Fig. 8.3. Spatial development of afforestation at Burrator Reservoir, 1900–1979.

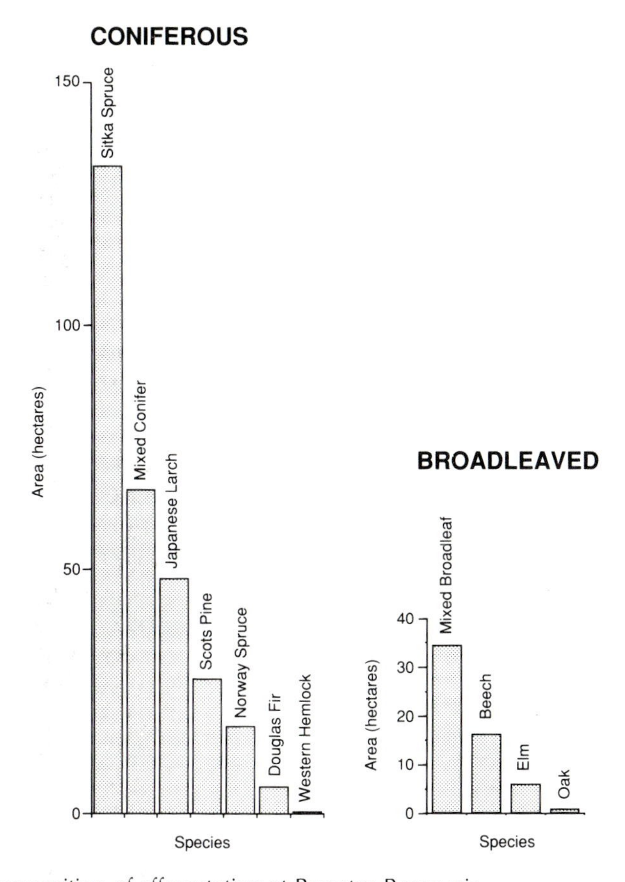

Fig. 8.4. Composition of afforestation at Burrator Reservoir.

the commercial need for a quick-growing crop. The most common species mixtures in the coniferous plantations were Sitka spruce (*Picea sitchensis*), mixed conifer (Sitka spruce and Douglas fir (*Pseudotsuga menziesii*)), Japanese larch (*Larix kaempferi*) and Scots pine (*Pinus sylvestris*). Sitka spruce was popular throughout the period, while mixed conifer, Japanese larch and Scots pine were predominantly used during the 1930s. All but four of the broadleaved stands (57.4 ha) in the plantations were original woodlands existing before the flooding of the valley. The broadleaved plantings occurred on the better soils and were largely composed of oak (*Quercus petraea*) and beech (*Fagus sylvatica*).

Development of the afforestation

The original afforestation plan by Plymouth Corporation was to plant all the enclosed agricultural land and to leave the open moorland unplanted. The

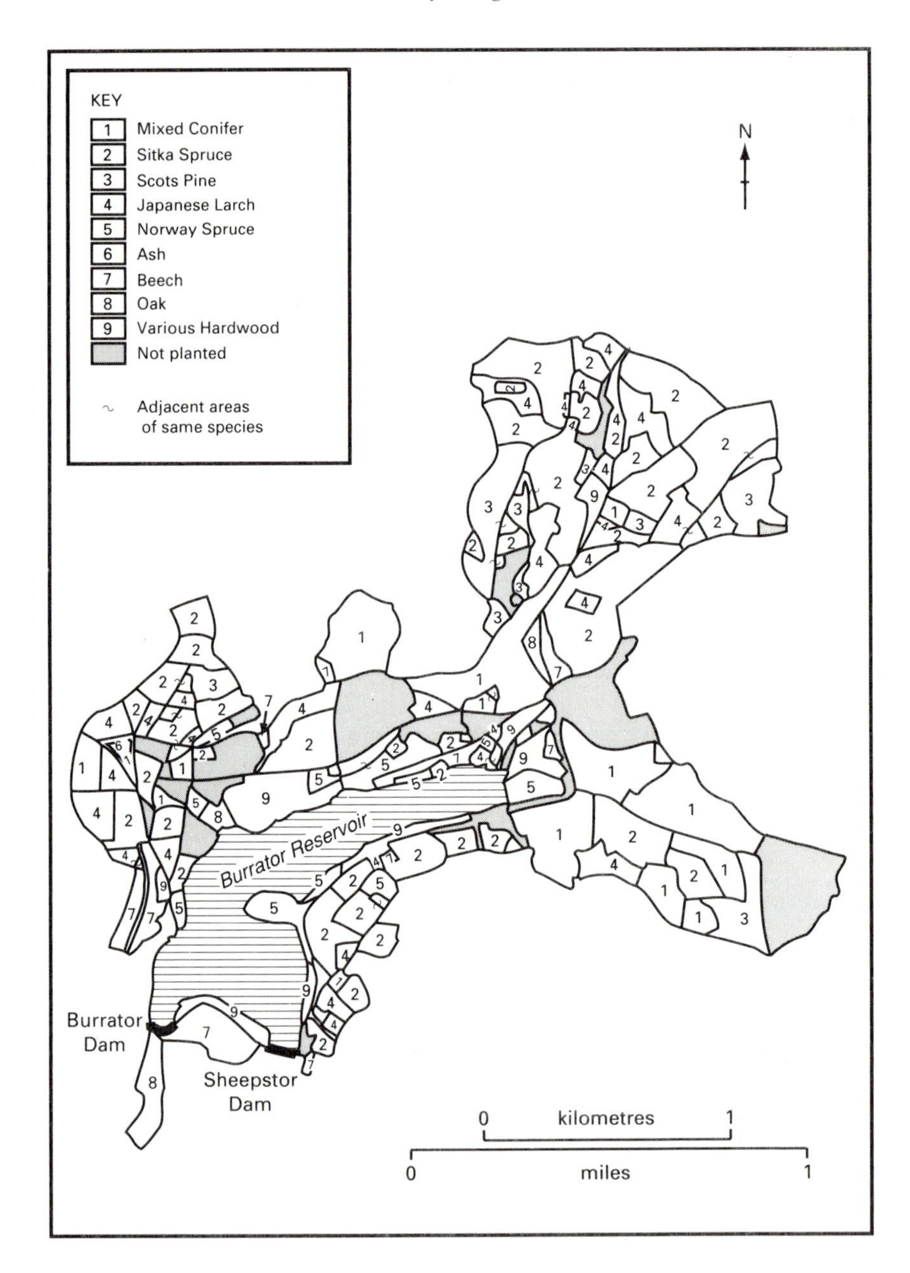

Fig. 8.5. Composition of afforestation at Burrator by compartments.

intention was to plant 12–16 ha/year (i.e. one compartment per year), which was considered feasible given the staff and finance available. However, the afforestation scheme was not on a large scale and developed in a piecemeal fashion. Furthermore, the workforce undertaking the afforestation at this time were workers from the water-management side of the operations, rather than trained foresters. The rate of planting in the inter-war years progressed at a steady rate and, by 1940, three-quarters of the present area of the plantations had been established (Fig. 8.6). The majority of the plantations were therefore reaching maturity between 1980 and 1991, which has important implications for the ecological evaluations undertaken in this study.

Since 1940, there has been a low level of management in the plantations, particularly in certain periods. During World War II, labour shortages caused a lull in the planting, water management took priority over forestry work and some of the catchment was required for the growing of potatoes and carrots for the war effort. The programme of first thinning, which became necessary during the war as many of the plantations reached 15–20 years of age, was also affected by these factors. No new planting occurred during the 1950s and the responsibility for afforestation continued to lie with labour from water management for much of the immediate post-war period. Plymouth Corporation did, on occasion, contract the Forestry Commission to undertake the thinning of trees, but the Corporation's commitment to the proper management of the afforestation over

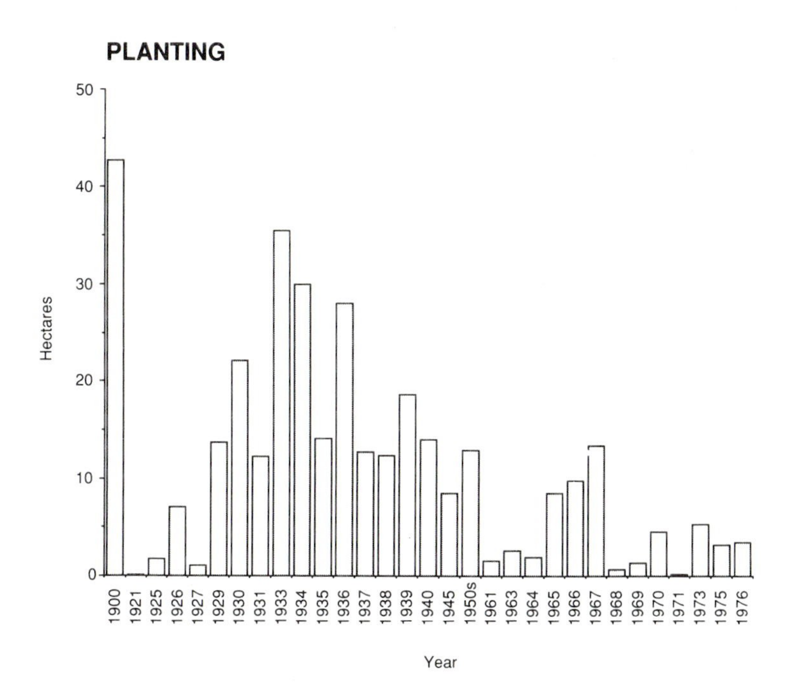

Fig. 8.6. Rate of planting at Burrator Reservoir.

this period was evidently low. Development of forestry around Burrator was also increasingly affected by the intervention of the Dartmoor National Park Committee, which had been formed in October 1951, and gave emphasis to landscape, nature conservation and recreation. Some areas around the reservoir were left unplanted due to such amenity pressures.

During the 1960s this situation began to change as Plymouth Corporation recognized the potential value of the timber from the thinning programme that was then necessary. The plantations required continual management to achieve good returns, rather than contracting the Forestry Commission to undertake the work at irregular intervals. Consequently, when the Water Manager/Forestry Officer retired in 1960, Plymouth Corporation took the opportunity to replace this post with a Forestry Officer (albeit with secondary duties concerning water management). The tasks set for the Forestry Officer during the 1960s were to undertake the thinning programme, to complete odd areas of replanting and to manage all plantations into a thoroughly efficient state of cultivation. A minimal amount of new planting was also undertaken, mainly consisting of Norway spruce to meet the market demand for Christmas trees.

Although these changes give an impression of a new commitment to the afforestation by Plymouth Corporation, in reality progress was hindered by the Corporation's financial structure. Revenue generated from the sale of timber was directed into the city's general fund, which was then not available to the Forestry Officer to fund planting and other management work. Such work could have been funded if the plantations around Burrator had been entered into a Forestry Commission Dedication Scheme, which would have paid annual management grants. However, Plymouth Corporation had refused to enter this scheme as a result of a misunderstanding about the consequences if targets set in the Dedication Management Plans were not achieved. Eventually, in the 1970s, the forests were entered into a Dedication Scheme agreement.

The variable commitment to management is also reflected in the level of employment at Burrator. During the 1950s, 18 men had been employed on forestry work. By 1961, this figure had fallen to 14 and by 1984 to eight. This decline in employment reflected the fact that much of the labour-intensive work of planting and thinning had been completed; that labour requirements had been reduced by the increasing mechanization of forestry work and that the long hours and low pay of forestry work were no longer attractive to potential employees. Subsequently, in anticipation of the sale of its plantations in 1985, South-west Water reduced the number of employees to just one part-time forester. Even though the sale was never completed, only one forester was employed throughout the late 1980s, and as a result any urgent forestry work had to be contracted out. Some of the work put out to contract during this period was too small to gain the interest of any contractors and so remained uncompleted. Such circumstances were particularly unfortunate since they occurred at a time when much of the afforestation was going beyond biological and economic maturity.

Summary of human influences

Several points emerge from this brief account of the development of the plantations at Burrator that might have some bearing on the ecological implications of afforestation. First, the rationale for the planting was a secondary consideration to the protection of the water supply for Plymouth. Coniferous species were selected for economic reasons and planting occurred on improved agricultural land. Second, the development of planting was piecemeal and not extensive, mainly occurring between 1925 and 1940 and covering only 16% of the catchment area. By 1991, therefore, the plantations were relatively long-established features of the landscape and ecology of the locality. Third, the intensity of management in the Burrator Forests had been, in general, fairly low. A relatively short period of afforestation by the non-forestry staff of Plymouth Corporation, during the 1920s and 1930s, was followed by a long period of general neglect. Even with the appointment of a Forestry Officer in 1960, the management input did not rise considerably due to labour shortages and institutional issues, such as the internal financing arrangements of Plymouth Corporation, the communication problems between Plymouth Corporation and the Forestry Commission over the details of Dedication Plans and the uncertainty concerning the possible sale of the plantations by South-west Water in the mid-1980s. The ecological consequences of these human circumstances will now be discussed.

Physical site description

In terms of the physical context, the Burrator Forests are located on the south-western margins of Dartmoor (see Fig. 8.2). Average annual rainfall ranges from 1400 mm to over 1800 mm and the elevation of the planted area lies between 220 m and 335 m, and slope angles generally lie between 5 and 10°. The entire area is underlain by granite. Soils on the valley sides, which is where the trees have been planted, are generally improved brown podzolics. These soils have a relatively high base status partly because of their topographic location and partly because of liming in the past. There is a long history of vegetation in the area; this had probably evolved by the Saxon era (800 AD) and has been grazed more or less extensively ever since (Gill, 1970).

The distribution of vegetation community types present in the Burrator area is closely related to natural and anthropogenic factors. The major vegetation communities of the Narrator catchment, one of the major Burrator watersheds, have been described in detail by Kent and Wathern (1980). The valley sides are dominated by upland *Festuca/Agrostis*, base-poor grassland with extensive invasion by bracken (*Pteridium aquilinum*). Such areas have a low floristic diversity and contain only about 20 species. There is a vegetation gradient down the hillsides, from the more acid-tolerant species at the top of the slope to the more neutral species at the base, that is related to soil characteristics. The upper slopes are characterized by a high proportion of *Festuca ovina* and *Vacinnium myrtillus*, while at the base of slopes *Agrostis capillaris* and herbs such as *Galium saxatile* and *Potentilla erecta* dominate.

Flora

Reconnaissance botanical surveys at Burrator have been conducted since 1977 to investigate the effect of crop species, management and species variation through the forestry rotation. The results from both recently forested and long-established plantation areas have been summarized in Table 8.1 and generalized into the scheme in Figure 8.7. Complete canopy closure was reached after 15 years, with consequent elimination of all plants at ground level. In the larch plantations, bryophytes were dominant after 30 years and grass began to colonize due to slightly enhanced light levels. No flora developed under the Norway spruce. However, about 45 years after planting, the flora re-established itself from the original seed bank.

The change in ground flora as the trees aged from 50 to 60 years was most dramatic. Litter dominated the 50-year-old Sitka spruce sites in 1977 but 14 years later there was no sign of the litter carpet and grass covered the forest floor.

Table 8.1. Botanical survey results for various tree crops through time.

Crop species	Age (years)	Management/ spacing (m)	Flora
Larch	15	2	100% litter
Norway spruce	30	2	100% litter
Larch	30	4	85% bryophytes, trace *Agrostis capillaris*
Sitka spruce	46	5	60% litter, 70% bryophytes
Sitka spruce	55	5	100% litter
Douglas fir	55	10	25% litter, 60% *Agrostis capillaris*, 25% *Galium saxatile*
Sitka spruce	60	5	50% ground cover of 11 species including 60% *Agrostis capillaris*
Sitka spruce	60	5	50% grasses including *Deschampsia flexuosa*
Larch	60	10	100% ground cover including bracken and brambles
Sitka spruce	70	5	100% ground cover, mainly grasses such as *Agrostis capillaris* and *Festuca ovina*; also *Oxalis*, brambles and *Vacinnium* present
Mixed broadleaved	2	3	About 10% bare ground; grasses, predominantly *Agrostsis* spp.; *Ranunculus* spp. and *Cirsium* spp. present; *Ulex europaeus* invading

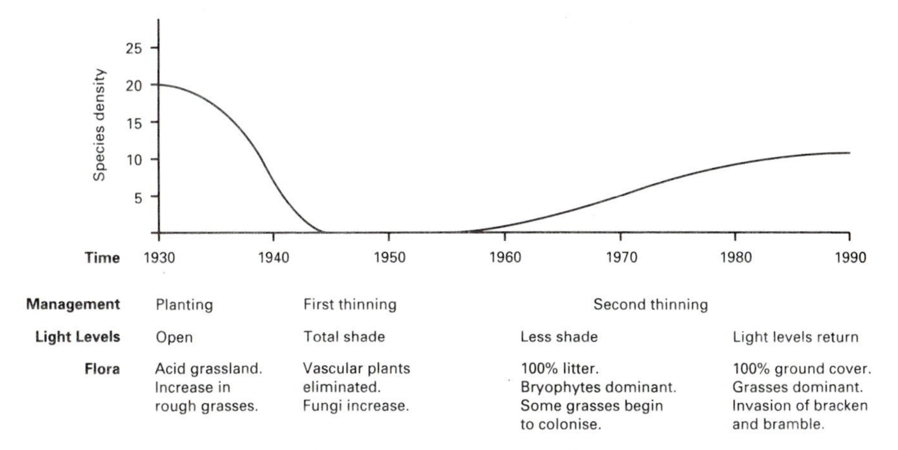

Fig. 8.7. Changes in ground flora through a forest cycle.

Agrostis spp. dominated but *Holcus mollis* was also present together with *Galium saxatile* and *Rumex acetosella.* In other areas *Agrostis capillaris, Deschampsia flexuosa, Endymion non scriptus, Rubus fruticosus* and *Pteridium aquilinum* were observed. Overall, the flora was species-deficient with only half the number of species observed in the grassland; nonetheless a true woodland habitat had developed. Although the initial impact of afforestation on existing flora was severe, subsequently a woodland ecosystem, which was more complicated in structure and function than the original acid grassland, was established beneath the conifers.

Several areas were examined following clear-cutting and replanting in 1990–1991. Although visually the sites looked very scarred with about 10% bare ground, the species composition was similar to the surrounding acid grassland. As well as the usual species, buttercups (*Ranunculus* spp.), thistles (*Cirsium* spp.) and foxgloves (*Digitalis purpurea*) were also observed. Gorse (*Ulex europaeus*) was also invading at several locations.

Soils

Thirteen soil profiles on the hill slopes in the Narrator watershed were described in the conifer plantation (five) and nearby grassland (eight) in order to determine whether there were any changes due to afforestation. Sample sites were selected along transects from the base of the hill slope to the junction with the moorland plateau.

The main differences observed between the forested and grassland soils were in terms of the litter layer, pH, roots and biological activity. In the soil profiles within the 50-year-old forest, there was usually 5 cm of litter in various states of decomposition, whereas in the grassland there was a turf mat. Soil pH was lowest in the forest. Organic acids released from the leaf litter resulted in the pH of the forest soil (4.0) being lower than that in the grassland (4.5). The number of fine and woody roots, as well as the amount of biological activity, was noted for each vegetation type. In the woodland, fine fibrous roots were common, with between 10 and 25 roots/cm^2, and there were three or four woody roots with diameters greater than 1 cm in an area 50 cm by 50 cm. In contrast, there were no woody roots in the grassland. The organic matter was better mixed in the grassland, which indicates more biological activity was occurring at this location.

Soil analyses were conducted to fractionate free iron and aluminium. Organic iron and aluminium were the dominant fractions in soil horizons from both the forest and the grassland, although they were greatest in the forest. Carbon contents were also high there. These soils exhibited high rates of organic sulphur formation, which may have implications for stream-water acidity.

Stream water

Stream water in the Narrator Brook was monitored daily for 1 year above and below the forest (February 1977–February 1978). Thereafter, water quality below the forest was monitored infrequently and, in April 1988, a programme of monitoring at monthly intervals commenced. The results for the year 1977–1978 are shown in Table 8.2 and are discharge weighted to take into account the different flows when sampling occurred. In addition, the forest had reached maturity so that its influence on water chemistry, if any, would be at its greatest. The results show that solute concentrations were greatest at the catchment exit with the marine-derived ions, sodium and chloride, showing the greatest gains. Acidity decreased slightly through the forest as did the calcium levels. However, no significant difference in the pH of stream water was found above and below the forest. This result is probably due to the catchment being dominated by brown podzolic soils, which, together with the great depth of weathered granite

Table 8.2. Discharge-weighted mean solute concentrations above the forest and below the forest (1977–1978).

	Above forest	Below forest
Specific electrical conductivity (μmhos/cm at 25°C)	50.1	55.0
pH	5.4	5.5
Na (mg/l)	6.6	7.1
K (mg/l)	0.9	1.0
Ca (mg/l)	1.1	1.4
Mg (mg/l)	0.9	1.0
SiO_2 (mg/l)	5.8	6.4
Cl (mg/l)	9.7	10.5

through which the groundwater passes, contain sufficient calcium to buffer the acid water.

Comparison between the forest exit water-quality results for the pre-cut (1977–1978; see Table 8.2) and post-cut (1988–1991) revealed few differences. The pH, for example, was almost constant at 5.6 as were sodium and chloride at 6.4 mg/l and 10.5 mg/l respectively. It is arguable that there is a lag in the system so that pre-cut conditions in the soil still prevail. The effects of the clear-cut will only be noted after many years as the soil adjusts to a new equilibrium. However, some of the soil characteristics have already changed and, for example, the monthly aluminium results have been declining steadily recently from 0.8 mg/l in 1988 to 0.3 mg/l in 1991.

Conclusions

The management of the coniferous afforestation at Burrator was only of a relatively low intensity; partly because of the lack of labour, development of the plantations proceeded in a piecemeal fashion and the rotation turned out to be about 70 years. In places, a distinctive woodland flora was able to develop. The forest soils were more acid, had high carbon and sulphur contents, but otherwise showed little physical or chemical differences to those from the surrounding open grassland. There was no evidence of increased podzolization under the forest, as might have been expected from evidence from similar sites elsewhere. The relatively high base status of the soils helped to buffer them against change.

The future is difficult to predict because there is no guarantee that past relationships between the human and physical factors will be sustained and because a new set of environmental conditions has been created. Any changes to the soil and vegetation due to re-afforestation will be superimposed on existing forest-type conditions.

The relationship between the human and physical systems at Burrator seems set to change as the full implications of the privatization of the Water Authorities are realized. The new water company now operates in a different financial environment and has introduced new forest policies, which are likely to initiate different ecological consequences in the area. Two policy changes are particularly significant in this respect. First, the woodland estate of South-west Water is now managed on behalf of the company by the Enteiprise Division, which is responsible for making a profit for the company and shareholders. Although a new management regime has been provoked by the wholesale devastation brought about by the hurricane-force winds in January 1990, the economic value of the afforestation has been highlighted once more, together with a realization of the potential value of the Burrator Forests as a recreational resource. Second, the new water company has introduced a general policy for its afforestation to plant broadleaves wherever practicable. It is proposed that within 50 years the composition of timber trees will be 85% hardwoods and 15% conifers. The hardwoods will mainly consist of oak, beech, ash, alder and willow, while cherry and guelder rose will be used for landscape reasons.

These new management policies will alter the present ecology. Light levels at the ground surface in the new planting will be increased for a considerable length of time due to the spacing, slow development and seasonal leaf pattern. Total shade is likely to only develop at the height of summer. Shade-tolerant grassland species will survive and ruderals plus some woodland colonizers are likely to become established. As the trees develop, the presence of so-called 'forest' species or stress-tolerators in the ground flora will increase through time. It is also likely that, as the forest develops, wildlife will increase and, for example, songbirds (such as blackbirds and stonechat), woodpeckers and buzzards will thrive (see Chapter 9). Deer are expected to come into the woods.

Broadleaved trees might benefit the soil in several ways. The nutrient cycle could be improved partly due to the presence of nutrient-rich leaves and deeper rooting. Leaf litter decomposition will be more rapid and nutrient cycling more active. Nitrogen mineralization rates will be enhanced. Water yield may increase in winter when interception is minimal and water quality will be maintained or may be improved.

The value of this research has been to establish a baseline against which future changes can be monitored. As the environment progresses from a coniferous to a broadleaved forest, changes in ground flora, the nutrient cycle of the soil and water quality will be monitored and related to management decisions. The dimensions of human-based research will focus on the rate of replanting, provision of recreational facilities and, in the long term, the economics of the new scheme. Water companies have a particular responsibility for large parts of the British uplands and should be aware of the possible environmental impacts of their changed forest policy.

ACKNOWLEDGEMENTS

The authors wish to thank South-west Water for their help and permission to conduct this research. In particular, Mr G. Taylor, the Woodland Ranger, is thanked for his interest and cooperation. The assistance of Professor Pinder, Drs Ternan and Kent, Miss A. Kelly, Mr B. Rogers, Mr T. Absalom and Mrs J. Sugden is gratefully acknowledged.

9

The Impact of Afforestation on Upland Birds in Britain

C. LAVERS AND R. HAINES-YOUNG

SUMMARY

Afforestation has been the most rapid agent of upland land-use change in Britain over the last decade. The area most severely affected by afforestation was the Flow Country of northern Scotland (Fig. 9.1), a region supporting the breeding populations of several species of internationally important birds (RSPB, 1985; Bainbridge *et al.*, 1987; Stroud *et al.*, 1987; Lindsay *et al.*, 1988). Conversion of land to forestry was so rapid, and knowledge of the ecology of the indigenous birds so poor, that accurate prediction of the impact of forestry proved impossible.

This chapter firstly identifies those bird species most at risk from upland afforestation by way of their distribution, ecology and rarity. In this way, an initial assessment of the likely importance of an upland area for birds can be obtained by a census of these key species. It goes on to assess the current status of theoretical ideas about the regulation of bird population processes that have been used to produce qualitative predictions of the impact of forestry. It is suggested that, as yet, there is little empirical justification for these ideas. Many more detailed empirical investigations of upland bird populations are required before the impact of forestry in a given region can be predicted with any confidence. Such studies should concentrate on the threatened species identified in this chapter in order to produce the best practical conservation benefit.

INTRODUCTION

The British uplands have developed a flora and fauna of considerable ecological and conservation significance (Ratcliffe, 1977; Hudson, 1988; Ratcliffe and

Thompson, 1988). In the last 30 years or so the uplands have come under increasing pressure from commercial afforestation. This chapter will concentrate on the birds of upland areas in Britain and the threat to their populations posed by upland forestry.

Beginning in the late 1970s, forestry began to expand in one of the most important ornithological areas in Britain and Europe: the Flow Country of northern Scotland. The importance of this region, however, was not fully appreciated until forestry had already affected large areas of upland habitat. Large-scale ornithological surveys, conducted initially by the Nature Conservancy Council, and followed by the Royal Society for the Protection of Birds (Stroud *et al.*, 1987), started at about the same time as forestry began to expand and, to some extent, forestry out-paced our knowledge of the true changes that were taking place.

To prevent such a situation recurring, information must be available not only on the bird species present in a forestry targeted area, but also on which of these species are likely to be most at risk because of their distribution, sensitivity to forestry, or rarity. With this information available, the ornithological importance of areas, and the way in which this importance is likely to be affected by forestry, can be assessed prior to any habitat disturbance.

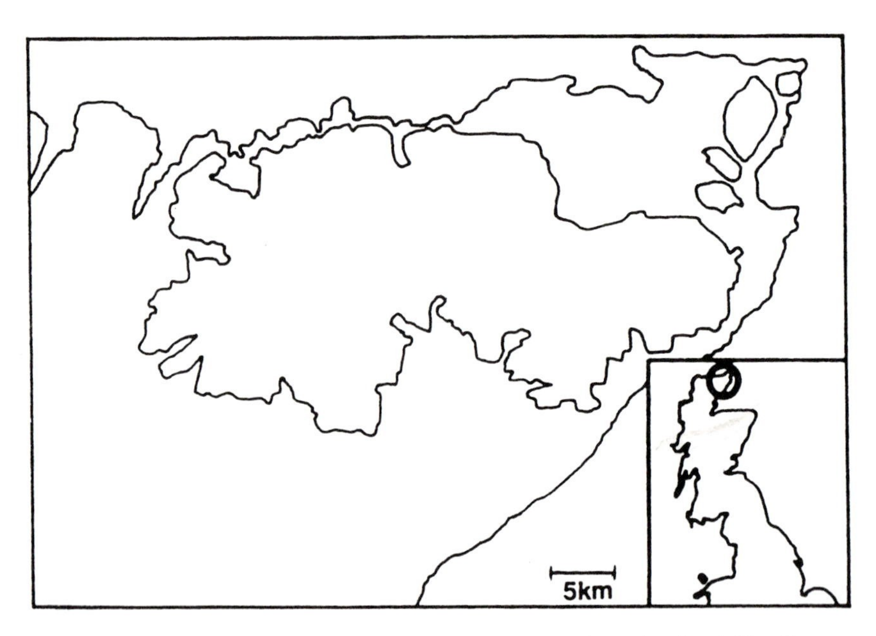

Fig. 9.1. Location of the Flow Country study area in northern Scotland. The five areas outlined contain relatively flat, sub-montane blanket peatland. These areas support significant proportions of the EC populations of dunlin, golden plover and greenshank. Nearly 20% of this area is currently in forestry ownership most of which was purchased over a short period after 1978.

When reviewing the literature, it is clear that little is known in detail about the effects of forestry on upland bird species. Broad principles are discussed, usually based around theoretical ideas, but few empirical studies exist. This is a serious situation, as many of the theoretical ideas on which generalizations about the effects of forestry are based, have come under increasing scrutiny in the last decade. Concepts such as resource limitation, habitat saturation and density-dependent processes (concepts that have a significant bearing on consideration of the effects of forestry) can no longer be advanced as evidence for a given point of view without empirical demonstration of their validity in the case of the species or communities concerned (Wiens, 1984; 1989; Lavers and Haines-Young, in preparation).

With these points in mind, the object of this review is threefold. Firstly, it will make use of an extensive list of upland birds (Thompson *et al.*, 1988) to identify those species that are likely to be affected by upland forestry. Many of the species identified, however, although common in the uplands, have extensive ranges outside upland areas and will consequently be less at risk than others. To identify and eliminate these, the most recent review of those birds whose ranges correspond most closely to upland areas will be employed (Avery and Leslie, 1990). The resulting species list will then be reduced further by considering details of the species' ecology or distribution within upland areas, that would tend to protect them to some extent from forestry expansion as it is likely to progress.

The second object of this review is to assess the importance of the species identified from the previous section, in both the British and European context. Again with the aim of directing research to the best conservation benefit, this will identify those birds that are most at risk because of their rarity.

Thirdly, current ideas about the effects of afforestation on upland birds are discussed. As mentioned previously, many of these revolve around theoretical considerations, rather than empirical investigations, and some of these theoretical ideas have been under increasing scrutiny in the last decade. There has also been a tendency for conclusions drawn from inadequately tested theoretical ideas to be dogmatically repeated, usually by organizations on either side of the forestry debate, without careful consideration of their current status. The first part of this section will outline the predicted effects of forestry and will then go on to consider the possible consequences for bird populations should these predictions prove not to apply in a given situation. The validity of previous attempts to produce quantitative estimates of forestry impact are assessed in this light. As a result of the lack of empirical studies, refutations of these classical ideas are as hard to find as corroborations, so much of this section will deal with possible consequences, although wherever possible the discussion will centre on the bird species identified as most at risk from the previous analysis. Results from several case studies are presented in conclusion, concentrating on the threatened species and theoretical ideas previously identified.

IDENTIFYING UPLAND SPECIES AT RISK

A list of 71 species associated with the uplands has been compiled by Thompson *et al.* (1988), based largely on the work of Ratcliffe (e.g. Ratcliffe, 1986). The complete list is reproduced in Table 9.1. To isolate those species most restricted to upland areas, the list can be compared with another of 31 species compiled by Avery and Leslie (1990). This list was constructed by an analysis of the *Atlas of Breeding Birds in Britain and Ireland* (Sharrock, 1976). The atlas provides information on the presence or absence of birds based on 10 km × 10 km squares of the National Grid and is the best information currently available on the breeding range of British birds; it also contains habitat overlays. The overlay showing the extent of moorland vegetation was used in the analysis, and those birds that had more than 50% of their records in moorland squares were identified. The birds representing the intersection of these two lists are recorded in Table 9.1 with minuses. This eliminates most of columns two and four of the original list, that is, those species that have extensive ranges outside upland areas. This does not necessarily mean that the populations of these species are safe from upland afforestation because the extent of a range says little about the importance of different habitat types within the range for maintaining the whole population. Similarly, conservation measures that aim to protect only the 'best' areas for a species are liable to lead to range fragmentation and the increasing isolation of metapopulations in similar habitats. However, the procedure outlined above gives some idea of the relative habitat specificity of the birds concerned and can be used, with qualification if necessary, to identify those species that are at reduced risk from upland afforestation.

The revised list (minuses) can be cross-referenced with the original list to identify those species with ranges largely restricted to the uplands and most at risk from afforestation. Fourteen species in Table 9.1 (marked '*') are 'immediately displaced by forestry developments or have their ranges substantially reduced by them' (Thompson *et al.*, 1988). Eight species 'most likely to decline through further forestry expansion' are marked with crosses.

The cross-referencing of these two lists eliminates those species that are undoubtedly upland birds but which are unlikely to be adversely affected by further increases in afforestation. These include species whose range lies mostly or wholly above the tree line (e.g. ptarmigan, dotterel), those who may benefit to some extent by afforestation (e.g. short-eared owl, black grouse), and some species that have close association with coastal regions (e.g. black-headed gull, great skua, red-breasted merganser) or whose habitat specificity would generally protect them from the effects of afforestation (e.g. goosanders in maturely wooded riparian corridors). The procedure also eliminates many of the upland species wholly or partially associated with moorland water bodies (e.g. common scoter, wigeon, greylag goose, common gull) as these are not considered to be so seriously at risk from afforestation as some of the open moorland species (Thompson *et al.*, 1988). There is some evidence for this (Haines, 1981;

Table 9.1. Bird species associated with upland areas.

1 **Upland specialists**	**2** **Upland/opportunistic**
Ptarmigan-	Peregrine-
Dotterel-	Raven-*
Snow bunting-	Buzzard-*
Red grouse-*	Kestrel
Golden plover-*	Red kite
Dunlin-*	Carrion/hooded crows
Twite -+	Meadow pipit
Whimbrel-	Skylark*
Red-necked phalarope	Wren
Arctic skua-+	Wheatear*
Great skua-	Whinchat
Greenshank-*	Stonechat+
Wood sandpiper-	Cuckoo
Temminck's stint	Lapwing*
Ring ouzel-*	Snipe*
Golden eagle-+	Redshank*
Merlin-*	Curlew*
Hen harrier-	Black grouse-
Short-eared owl-	Chough+
3 **Use upland water bodies**	**4** **Generalists with upland distributions**
Black-throated diver-+	Jackdaw
Red-throated diver-+	Tree pipit
Wigeon-	Grasshopper warbler
Goosander-	Pied wagtail
Red-breasted merganser-	Mistle thrush
Common scoter-	Song thrush
Teal	Oystercatcher
Dipper+	Ringed plover
Grey wagtail	Herring gull
Greylag goose-	Lesser black-backed gull
Common sandpiper	Great black-backed gull
Common gull-	Stock dove
Mallard	Nightjar
Black-headed gull-	Whitethroat
Slavonian grebe	Tawny owl
	Willow warbler
	Redwing
	Fieldfare

Source: Thompson *et al.*, 1988.
See text for explanation of symbols and appendix, p.151, for scientific names of the species

Eriksson, 1984) but, as Ratcliffe (1986) points out, the effects of afforestation on these birds may be indirect but they may be potentially serious; as yet we simply do not know enough about these indirect effects to make a definitive judgement. The British populations of common scoter and greylag goose especially, because of their distribution and ecology (Avery and Leslie, 1990), may be particularly at risk.

By the procedure outlined above, the 22 species most likely to be affected by afforestation, and the 31 most characteristic of the uplands, have been reduced to the 13 upland specialists most likely to be seriously affected by the current rapid rise of upland forestry (Tables 9.2 and 9.3). The birds most likely to be affected are greenshank, golden plover, dunlin, raven, buzzard, merlin, red grouse and ring ouzel from the 'most-easily-displaced' group, and golden eagle, twite, arctic skua and the divers from the second group, which are 'less-at-risk', or less known to be at risk.

The members of the second group do not qualify for membership of the first group for various reasons. The divers, for instance, are unlikely to be 'displaced' in the same way as open-moorland species, as forestry avoids their required habitat (lochs). However, forestry in many instances has tended to enclose lochs completely (divers nest on the loch edges) and the trees may also obscure the flight path of the birds – especially where lochs are small. Changes in run-off characteristics may also lead to rapid fluctuations in water levels in lochs, which

Table 9.2. Selected references to bird species threatened by British upland forestry.

Bird species	References
Golden eagle	Marquiss *et al.* (1985), Watson *et al.* (1987).
Merlin	Newton *et al.* (1978; 1984; 1986), Bibby (1986; 1987).
Raven	Marquiss *et al.* (1978), Newton *et al.* (1982).
Buzzard	Picozzi and Weir (1974; 1976), Newton *et al.* (1982).
Greenshank	Nethersole-Thompson and Nethersole-Thompson (1979), Nethersole-Thompson and Watson (1981), Thompson *et al.* (1986).
Golden plover	Ratcliffe (1976), Reed *et al.* (1983).
Dunlin	Soikkeli (1967; 1970), Holmes (1970), Avery and Haines-Young (1990).
Red grouse	Miller *et al.* (1966), Jenkins *et al.* (1967), Lance (1978), Watson and Moss (1980), Watson *et al.* (1984).
Black-throated diver	Bundy (1979), Campbell and Talbot (1987).
Red-throated diver	Gomersall *et al.* (1984), Thom (1986).
Ring ouzel	Flegg and Glue (1975).
Twite	Orford (1973), Davies (1988).
Arctic skua	O'Donald (1983).

Table 9.3. Importance of bird species threatened by British upland forestry. Species are listed in order of the proportion of their range occurring in moorland grid squares (Avery and Leslie, 1990). For details of the British and European directives (columns 1–3) see Stroud *et al.* (1987).

	1	2	3	4	5	6
Greenshank	*	*		66	66	254
Red-throated diver	*	*	*	14	14	212
Black-throated diver	*	*	*	20	20	319
Golden plover	*			18	17	915
Ring ouzel	*					780
Raven						1697
Twite	*		*			785
Dunlin	*		*	39	35	537
Red grouse[a]						1503
Merlin	*	*	*	5	4	843
Arctic skua	*			2	2	139
Golden eagle[b]	*	*				108
Buzzard						1451

[a]Subspecies *scoticus* exists elsewhere only in Ireland.
[b]Britain holds all of the nominate race, 6% of which are in Caithness and Sutherland; including the southern race *homeyeri*, Britain holds $<1\%$ of the EC population.

Column 1 = annex 1 species of EEC birds directive; column 2 = Schedule 1 species of Wildlife and Countryside Act; column 3 = Appendix 1 species of Bern Convention; column 4 = % of British population in Caithness and Sutherland, N. Scotland; column 5 = % of EC population in Caithness and Sutherland; figures where available; colum 6 = number of 10km × 10km grid squares in the *Atlas of Breeding Birds in Britain and Ireland* (Sharrock, 1976) occupied by each species; see text.

may in turn lead to nest failure by flooding, so divers are by no means safe from the immediate proximal effects of afforestation. Red-throated diver and arctic skua have their national population strongholds north of mainland Britain (mainly Shetland). This may protect them to some extent as such areas are not currently targeted for forestry. The distribution of twite corresponds mainly to moorland fringe areas, and this would tend to protect it to some extent from forestry if the current pattern of planting persists. The effect of afforestation on golden eagle is difficult to predict (but see Marquiss *et al.*, 1985; Watson *et al.*, 1987) as this species requires such large territories: it would have to be a considerable forestry development that covered the whole territory of a golden eagle, although this is of course possible. The extent to which golden eagles can suffer losses of their territory to forestry is not known in detail, and, like ravens, will probably vary depending on local circumstances.

The presence of buzzard in the 'most-at-risk' group, is probably wrong. The risk to the buzzard is no greater (probably less so) than that to the golden eagle for instance. Afforestation may benefit buzzard by increasing the density of small mammals in the early years after planting (Newton *et al.*, 1982; Ratcliffe 1986), or by increasing nest site availability as the forest matures. These factors, combined with a decrease in persecution from gamekeepers in the vicinity of plantations, could well lead to increases in buzzard numbers. For these reasons, we consider that buzzard should be in the 'less-at-risk' category.

There are some notable omissions from the final list presented in Table 9.3. Osprey and red kite are probably omitted from the list of upland specialists because their range in the atlas is deliberately plotted incorrectly to protect their nest sites. These two species are not particularly sensitive to limited afforestation, however, and may benefit to some extent (Newton *et al.*, 1982; Thompson *et al.*, 1988; Avery and Leslie, 1990). Other species that many would include as upland species (e.g. common sandpiper, curlew and dipper) have ranges that are too well represented in lowland habitats to be included (Avery and Leslie, 1990). Some rare species are probably not included as the probability of finding them in any atlas square is so low (e.g. red-necked phalarope, Temminck's stint, wood sandpiper and others). An open mind should also be kept about the species associated with upland water bodies, in particular common scoter and greylag goose, as any adverse effects of forestry, however indirect, could have serious consequences for the viability of the populations of these species.

Importance of British Upland Species at Risk

How important are the 13 bird species identified as being most seriously threatened by upland afforestation? This depends to a large extent on the criteria used in the assessment. Some points about this are outlined below. However, to give some idea of the threat that upland forestry in Britain poses for birds that have importance in the context of Britain and the European Community (EC), Table 9.3 summarizes the status of these species under several National and European Acts, Directives and Conventions concerned with the protection of important and endangered wildlife and their habitats. Estimates are also given of their status in the rest of the EC, and the proportion of the British population held in the counties of Caithness and Sutherland (the 'Flow Country'; table adapted from Stroud *et al.*, 1987). The number of 10 km × 10 km National Grid squares in which each bird is recorded in the atlas has also been added to give some idea of the relative ranges of each species and the availability of suitable habitat in Britain (from Avery and Leslie, 1990).

From Table 9.3, the extent of the problem concerning the spread of upland afforestation in Britain appears particularly serious. Many of the species most at risk from such developments (especially the waders and divers) have their entire EC range, or a significant proportion of it, in Britain. Greenshank, for example, are estimated to have only 25 400 km^2 of suitable breeding habitat in the EC, and 66% of the EC population of this bird breeds in one small area of Scotland covering the peatland areas of Caithness and Sutherland. By the end of 1991,

16.2% of this area had been lost to conifers, 11.5% since 1979 (Lavers and Haines-Young, in prep).

These arguments about the importance of British upland populations need to be qualified. The use of the EC population of a given upland species in judging the importance of British populations is misleading. Conservationists have been quick to use EC-based criteria for promoting the importance of British upland birds without taking a wider perspective. In some cases this may be due to the lack of detailed information on population numbers outside the EC, but in other cases, EC populations are probably quoted because the EC as a whole is a very poor region for the British birds under consideration. The EC does contain other large upland areas (e.g. in France, Spain, Greece and Italy) but, due to their latitude and climate, these areas are very different from those in Britain. As such, they are unlikely to contain similar bird populations.

Scandinavia, Iceland and Russia on the other hand, have very similar tundra-like ecosystems and support similar species. The greenshank, for example, which became the 'emblem' of the battle between conservationists and foresters in the Flow Country, does not breed anywhere else in the EC. The importance of the Flow Country then, with 66% of the EC population, appears considerable. However, if Norway, Finland and Sweden were to join the EC, and introduce their estimated 97000 pairs of greenshank (Avery and Leslie, 1990), the importance of the Flow Country for greenshank all but disappears. The populations of greenshank in European and non-European Russia are probably even larger. Similar calculations are relevant for dunlin.

Avery and Leslie (1990) conclude that only the British populations of golden eagle, twite, red grouse and golden plover from the list of threatened species in Table 9.3 are of justifiable importance when Britain is considered in the context of the western Palearctic. Of these, red grouse and golden plover are important because the British races of these species are distinct and, if the northern and southern races of the golden eagle are combined, Britain holds less than 1% of the EC population of these birds (Stroud *et al.*, 1987). The twite in Europe breeds only in Britain and Ireland with a subspecies of *Carduelis flavirostris* breeding in coastal Norway and Finland (Thom, 1986). Twite also breed in the Himalayas, and the European birds are thought to be an isolated relict population from the last ice age (Thom, 1986). It could be argued with some justification that the twite, the definitive small brown passerine and one of the least studied and most easily overlooked of species, is the most important of Britain's open country birds when considered in the context of the rest of Europe.

The most appropriate way to assess the importance of British upland species under threat, therefore, is problematic. In going from national-based to global assessments of the importance of an area of bird habitat, the availability and accuracy of data on bird population numbers diminishes but the assessments so produced are more rational. Undoubtedly EC-based criteria are probably the most artificial and unsatisfactory from the ecological point of view, as the entry of, for instance, Sweden, into the EC would abruptly alter the conservation status of many of Britain's upland birds. Nevertheless, Britain's membership of the EC puts the UK government under obligation to protect certain species and their habitats that the EC recognizes as important. EC-based assessments of the importance of an area or country have some meaning, if only political. National-based criteria for judging the conservation importance of an area such as the Flow Country have similar political weight, but for a migratory bird that breeds in Scotland, winters in Africa and passes over Europe twice a year on migration, the importance of Scotland is only part of the story. There is no obvious solution to the problem of assessing the ecological value of an area for birds, and as such, all assessments need to be considered critically.

In the following discussion on the predicted effects of afforestation, examples from these species considered most at risk will be used wherever possible, or where discussion revolves around theoretical considerations, examples will concentrate on the likely consequences of afforestation in the habitats typical of these species.

EFFECTS OF AFFORESTATION ON OPEN COUNTRY BIRDS

It has been widely assumed that afforestation and population decline are linked in some simple way. This assumption is a key one, and needs to be examined critically. Research in recent years, especially in the charged atmosphere during the episode of forestry in the Flow Country, has tended to become politicized, if not by the researchers themselves, then by organizations and individuals on either side of the forestry debate. When reviewing the literature on the effects of afforestation on birds, it is noticeable how little explanation of the assumptions on which the research is based is given – especially when quantitative estimates of forestry impact are presented. Qualitative statements and broad generalizations about the relationship between forestry and open country birds seem reasonable, assuming the validity of ecological principles such as habitat saturation, resource limitation, competitive interaction as a determinant of population numbers and community structure. These principles, however, have become increasingly questioned over the last decade (see Lavers and Haines-Young, in preparation; and especially Wiens, 1984; 1989). Quantitative estimates are particularly sensitive to the validity of assumptions, and it is necessary therefore to examine the consequences if the assumptions are not satisfied. Two ecological processes predicted to be initiated or enhanced by afforestation, displacement and density-dependent population regulation, will be examined.

Direct effects

DISPLACEMENT

What is the response of dispossessed birds to habitat reduction or physical displacement by forestry? Do the birds die, or become non-breeders, or try to establish somewhere else? Is it possible for individuals of a given species to re-establish, if not in the year of displacement, then in subsequent years? If re-establishment is possible, where is this likely to take place and what (quantitatively) are the consequences for the receiver population? Empirical answers to most of these questions are unavailable. If birds have the flexibility to respond to forestry by trying to establish somewhere else, the general effect of displacement will be an increase in density of birds on the moorland that remains, and some restriction of previously available resources (forestry may add alternative utilizable resources to compensate). The effect of this constriction will probably vary from species to species, depending on which resource is limiting to the population (if any are in the area of afforestation), and also in what proportion of forestry area this restriction occurs. It will also depend on the distance of displacement. If birds are displaced onto ground immediately adjacent to the plantation, this would tend to produce narrow, high-density zones of birds. This may initiate or exacerbate density-dependent processes such as competition or predation. The greater the distance of dispersal, the less pronounced these effects are likely to be.

The restriction of resources through displacement may be different for different bird species depending on their resource utilization characteristics. An example of forestry restricting resources in proportion to area would be a herbivorous bird whose limiting food resource is moorland vegetation. 'Over-restriction' of resources per unit area of forest may occur for the same species, if forestry alters vegetational patterns beyond the boundaries of the plantation. 'Under-restriction' may occur if a bird species can occupy a zone within the edge of the forest but is excluded from the interior. Under-restriction may also occur if the favoured upland habitat of a particular bird is unplantable. There is evidence that under-restriction of territories is the norm in the Flow Country for dunlin. It has been shown (Lavers and Haines-Young, in prep) that forestry in this area avoids wet marshy areas and pool systems. By combining data on the distribution of pool systems with forest boundaries in a spatial database, it has been possible to assess the actual impact of forestry on pool systems (the favoured habitat of dunlin). It is also possible to model the hypothetical effect of a different forestry regime by artificially shifting the forest boundaries by a given amount in any direction. The results show that the number of pool systems actually affected is generally half of what would be expected under any other regime. Thus, hypothetically, 20% forestry coverage of an area may only result in 10% displacement. For other reasons this is not the case with dunlin (Lavers and Haines-Young, in prep).

In general, displacement and increase in density can be considered a bad thing for the population. This is so even if the population is not resource-limited, and can suffer variations in the number of birds occupying the breeding grounds without a proportional decrease in overall breeding success. The negative impact occurs because any limitation or density-dependent regulation of population numbers that might occur, even if only sporadically, will tend to be exacerbated by density increases due to displacement. This much is virtually certain, but the detailed effects and the many factors that would complicate the quantitative estimation of these effects such as territoriality, philopatry (fidelity to sites of birth or previous nesting areas), whether limiting resources are area-linked, whether populations are resource limited in the particular area where afforestation is taking place and so forth, are poorly understood in all upland species with the possible exception of red grouse (Lance and Lawton, 1990).

The supposed effects of increasing density can be summarized with reference to the theoretical work of Svardson (1949), expanded and developed by Brown (1969) and Fretwell and Lucas (1969). These authors developed models to explain the likely mechanisms of range expansion, one of the possible consequences of displacement that is rarely acknowledged. Fretwell and Lucas (1969) explained habitat occupancy in terms of the fitness potential of habitats. Some habitats (or areas with certain habitat features) are of better 'quality' than others, and will tend to be occupied first. As density increases, however, intraspecific competition would be expected to increase, and the fitness potential of the habitat as a whole, to decrease. This continues until population density in the first habitat reduces its fitness potential to the point where the potential of a more marginal habitat is equal to the first, and both habitats are occupied. If the best habitat is occupied first, or by the fittest individuals to the exclusion of others, increases in density will lead to marginalization of any overspill. Marginal habitats, by definition, are less productive than habitats with higher fitness potentials, and densities and breeding success would be expected to be correspondingly lower. This model has been shown to have some validity for density increases in bird populations (see for example O'Connor, 1980; 1985; 1986; O'Connor and Fuller, 1985).

In recent years, however, questions have been asked about the role of density-dependent processes in regulating bird populations. Some of this research is particularly relevant to the question of whether population decline due to displacement by forestry will be predictable.

Density-dependent population regulation

The importance of density-dependent regulation has more often been assumed than demonstrated. The assumption of resource limitation in particular has been identified as one of the factors that contributed greatly to the stagnation of ideas in ecology prior to the 1980s (Wiens, 1989). Our lack of knowledge about the relationship between population density and population production gives ample scope for departures from the classical model of density increases outlined above.

In an extensive review of the shorebird literature, Goss-Custard (1981) found no evidence that density-dependent regulation had any effect on the number of chicks hatched or fledged per breeding pair for any of the species considered. He qualified this by saying that the effects of territoriality may result in production varying in an inversely density-dependent way, when density was expressed in terms of breeding plus non-breeding birds. Evans and Pienkowski (1984) question this assertion, however, using several lines of evidence. Territoriality may result in density-dependent regulation, even under circumstances of average densities of birds, provided that some birds do not breed. The evidence for this is based largely on the presence of these non-breeders in the population. It has to be demonstrated, however, that these birds are capable of breeding. This is rarely achieved (but see Harris, 1970; Holmes, 1970). The inducement of density-dependent regulation through territoriality also assumes that all viable habitat is occupied so that range changes are not possible. It would be particularly important to know whether displacement invariably results in the loss of birds from the reproductive population, or whether this can be offset by range expansion into suboptimal habitats.

It has been demonstrated (e.g. Soikkeli, 1967; 1970) that territory size and total breeding densities of dunlin in Finland are affected by the number of young produced 2 years previously, and that territory size is reduced depending on the number of birds that try to settle. Similar results have been recorded for greenshank (Thompson *et al.*, 1986, and see below). It is possible, even in a situation where territory size is a function of the total influx of birds, that there could be weak density-dependent restriction on average production per adult in the area, provided that some birds are prevented from gaining a territory (Goss-Custard, 1981). The significance of such weak regulatory processes is questionable, however (Evans and Pienkowski, 1984). It is also not uncommon for there to be a surplus of male birds that secure territories but do not secure females. This has been recorded for dunlin in Finland (Soikkeli, 1967). In such cases, it is unlikely that average female productivity is restricted by territorial behaviour. Evans and Pienkowski (1984) reach several conclusions that are worth repeating in full. For instance, that:

> there is as yet no firm evidence of density-dependent restriction of reproductive output in those species breeding at higher latitudes: that all females that are capable of breeding, hormonally and nutritionally, are able to find a mate somewhere within the distributional range of the species.
>
> (p.114)

and that:

> average reproductive output per female [of the species reviewed] varies from year to year and from place to place, depending in particular on local

> variations in the intensity of predation. There is no evidence that such variations are determined by population density.
>
> (p.114)

These authors conclude by wondering whether:

> the concept of population regulation in shorebirds is illusory; whether severe weather acting unpredictably to cause mortality on the wintering grounds and on the breeding grounds, where it may also prevent reproduction, serves to prevent shorebird populations from reaching the utopian levels of the carrying capacity of the habitats to which they are adapted.
>
> (p.117).

Similarly, Senner and Marshall (1984) in their review of the conservation of the wintering grounds of shorebirds, conclude that the effect of habitat loss is to displace birds that had used this habitat, that it can be assumed that displacement will not be beneficial but that:

> the ultimate significance of such displacement is largely a matter of speculation and will vary with specific circumstances.
>
> (p.400)

The implication in the present context is that if the carrying capacity of breeding grounds subject to forestry can support 'utopian' levels of birds, levels that due to factors not associated with bird density are rarely approached, then in terms of resource availability, the area would be able to support the reproductive effort of a higher density of (displaced) birds with some gain in population productivity. Winter mortality, for example, if severe under normal circumstances, may result in the general availability of usable habitat on the breeding grounds to a greater density of birds. The gain in population productivity is perhaps unlikely to equal the productivity had displacement not occurred (Thompson *et al.*, 1986), but the possibility of consistent gain at any level brings into question whether some of the methods of assessing population decline due to forestry are strictly valid without qualification (Stroud *et al.*, 1987; Avery and Haines-Young, 1990).

One can introduce further complications into the 'classical' model of the effects of density increase by considering the tenacity of some birds to particular breeding areas (natal philopatry or nest-site fidelity). As Wiens (1989) points out, site-tenacious behaviour can lead to situations where one of the main assumptions of the classical models of Fretwell and Lucas (1969) and Brown (1969), namely that the best habitats are occupied first, is violated. If, for instance, winter mortality in a given year is severe, bird density on breeding grounds would be low and unoccupied habitat may be available for occupation by individuals displaced by forestry. If the fitness potential of the habitat that has been lost to forestry is lower than that normally available, displaced birds may encounter territories with higher fitness potentials than those to which they were tied, even if they are latecomers into a new area. This is because the returning birds do not occupy the best habitat available to them due to site-tenacious behaviour. If displaced birds are forced to look elsewhere by displacement, then in this hypothetical situation (and in this year), population production when expressed per breeding pair may be higher as a direct result of displacement.

Of course many factors are assumed here, not least of which is the assumption that the tenacious behaviour that applies to the non-displaced population will not apply to the displaced birds. Species that exhibit strong site tenacity may be much less likely to re-establish elsewhere than birds that do not. Also, as Thompson *et al.* (1988) point out, there are various consequences of tenacious behaviour that may have adverse consequences for the population even if birds do attempt to re-establish. For instance, displaced birds may not encounter mates from the previous year, or they may be at a disadvantage through being unfamiliar with the new breeding area. Laying may be delayed for displaced birds, which may lead to reduced breeding success, and this in turn may lead to an increased divorce rate in subsequent years. Judging the effect of tenacious behaviour on population processes is complex.

The points outlined in this section give rise to doubts about whether population decline in some situations will keep pace with forestry development. It is probable, however, that in some circumstances decline will occur much faster than forestry advances. Population 'crashes' may occur if forestry limits the resources available to a population beyond a particular point. Each individual in a population can acquire enough of a non-limiting resource for normal productivity until it is reduced past a threshold, at which point no individual can acquire enough, and population production falls rapidly. Such considerations may be particularly applicable to birds with large territories, only parts of which become afforested, for example golden eagle, merlin, red kite, raven (Avery and Leslie, 1990), but it may apply to any species, such as greenshank (Nethersole-Thompson and Watson, 1981). Other hypothetical routes to 'supraproportional' population decline relative to forest area are presented by Thompson *et al.* (1988).

The consequences of various aspects of social behaviour and social structure on post-displacement population processes are complex, and the way in which populations are affected by disruptive influences to such behaviour therefore will tend to be similarly complex. Such considerations give rise to doubt about whether generalizations about population decline due to forestry will have very general application. At least this will be the case until much more is known about the response of upland birds to displacement, and the prevalence of density-dependent processes in upland communities.

EXAMPLES FROM THREATENED BIRD SPECIES

Some of the theoretical considerations mentioned above are now illustrated and expanded by considering several case studies of work on the threatened bird species identified in the first section. Few studies have investigated directly the effects of afforestation on upland bird species but those that have, as predicted, have yielded complex results. This is certainly the case for the first two studies on raven populations in different parts of Britain. The third study on greenshank does not concern forestry directly but the results suggest that, for the population studied at least, population production is not regulated by density-dependent processes in any simple way. These studies suggest that, in some circumstances, forestry may have less effect on a population than is quantifiable on the basis of the proportion of a region lost to forestry. In the final section, several case studies on 'edge effects', the most likely mechanism by which forestry could have a greater effect than that predictable from forested area alone, will be reviewed.

RAVEN

One example of forestry restricting available area but not reducing essential resources to the point where they are limiting is the work on buzzard and raven in agricultural and forestry areas in Wales (Newton *et al.*, 1982). For raven, one of the species most restricted to upland and coastal areas and most vulnerable to forestry expansion (Thompson *et al.*, 1988), no decline in occupancy or breeding success was observed in relation to total closed plantation cover (the birds did not utilize the closed forest areas). Nor did the distribution of birds suggest that a decline in numbers had occurred historically in forested areas as pairs were evenly spread regardless of plantation cover. The proportion of closed forest within 1 km of nests varied from 0–80%, and within 3 km of nests from 0–60%. From these data it appears that ravens in this area can lose (and have lost) up to 80% of their previous foraging range and suffer little or no adverse consequences. These findings suggest that forestry in this area, while restricting the area available for use, does not restrict food or any other resource to the point where it is limiting, and that the same number of birds are in occupation as were present prior to forestry.

Although no significant difference in the breeding success of raven was observed in relation to the proportion of a territory afforested, there was some indication that the amount of available carrion correlated with breeding success in the early part of the breeding cycle. Only 4.2% of the study area was considered in this part of the analysis, however; also the two areas in which carrion counts were made were at the extreme southern and northern ends of the study area. This spanned the greatest altitude range possible, so the possibility of other confounding factors influencing the result was maximized. The authors state that high nest occupancy and the presence of non-breeders suggests that density was close to what the area could support. However, it was shown that territorial occupancy had not declined through the loss of large parts of the study area to forestry; it is likely, therefore, that the same reasoning could have led to the same conclusion if the study had been carried out prior to forestry and prior to the loss of resources that this implies.

If food-resource limitation controls raven populations in Wales, one might expect that the low impact of forestry on raven numbers means that forestry adds food resources (e.g. by increasing live prey) that are lost under the closed forest. However, the finding that there was no significant difference in the composition of prey in pellets between forested and unforested areas, despite large differences in the availability of prey items (particularly carrion, which figured in 92% of the pellets in the breeding season and 86% outside), suggests that the birds in forested areas did not change their diet significantly to track differences in relative resource availability, and that the birds in unforested areas may not saturate the carrion supply, which is the most important food resource for ravens both here, and throughout the birds' range (Newton *et al.*, 1982). Some other resource may have been the limiting factor, or food may have been limiting only at one time of year. Or perhaps raven density is a function of resource availability during times of 'environmental crunch' (Wiens, 1989). However, the data are not available to test between these alternatives. Virtually the same results were obtained for buzzard in the same study, another of the threatened species identified earlier.

Markedly different results were obtained for ravens in Scotland and northern England (Marquiss *et al.*, 1978). Here both occupancy and breeding success were strongly inversely correlated with afforestation. The same species affected by forestry, therefore, can respond differently in different areas. The exact reasons for this are unclear, but the suggested reasons (Marquiss *et al.*, 1978; Newton *et al.*, 1982) generally employ untested inferences about food resources or uncontrolled comparisons of availability in different areas. Both studies assume resource limitation. For example, in the study of Marquiss *et al.* (1978), the summary contains three untested assumptions about food resources, which are used to explain the correlation of breeding density with altitude, the level of afforestation that an area could sustain before desertion and the desertion of marginal areas. Regarding the first inference, in the Welsh study, attempts were made to measure the extent of the principal food resource (carrion) and it was found that nest spacing was actually wider on higher ground where carrion was more abundant, the opposite of what was expected.

GREENSHANK

Greenshank is one of the species from our threatened list that is likely to be physically displaced into an adjacent area through loss of whole territories. These birds generally defend multiple territories for different activities, so loss of a feeding territory need not imply loss of a nesting territory. A 19-year study of greenshank populations in north-west Scotland (Thompson *et al.*, 1986) revealed some suggestive information about the effects of density. These authors found that the main constraints on breeding success were various aspects of the weather. Greenshank laid earlier and produced heavier clutches when early spring was warm. Hatching success and chick weight were all reduced when conditions during incubation were wet and cold. Fewer birds returned to the breeding grounds and fewer new territories were established when the previous 2 years had been characterized by harsh weather. Early laying was positively correlated with the presence of a river in the territory, and female age. Individual differences between females also had a consistent effect on laying date.

Differences in density between years was marked. The studied population is a small one, however, and one of the populations at the extreme of its breeding range (Thompson *et al.*, 1986), so the density variations recorded may not be representative of the species. Also, year-to year variations in the density of breeding waders in general is thought to be relatively low (Evans and Pienkowski, 1984; Langslow and Reed, 1985).

The number of breeding females in the study area averaged 15 over 19 years (one female per 100 ha). The mean difference in number in any one year compared with the previous year averaged 2.78 females, or 18.5% of the average population. Thus density differences from year to year varied by a factor approaching one-fifth of the average density. The minimum difference in number was zero, while the maximum was 7, or 47% of the average number. No clear underlying pattern of yearly density variations was obvious.

Part of the study attempted to ascertain whether increased density of breeding birds was reflected in the productivity of the population. If increased density had resulted in increased competition and this competition disproportionately lowered chick production, this would be good evidence that competitive intensity was density-dependent. There was a strong inverse correlation between fresh clutch weight and breeding density suggesting that some density-dependent factor may have been operating. However, the authors also defined a measure that reflected the viability of a given clutch and the potential breeding success (PBS) of a given female:

$$\mathrm{PBS} = \mathrm{C} \times \mathrm{H}$$

and also for the population as a whole:

$$\mathrm{PBS'} = \mathrm{F}(\mathrm{C'} \times \mathrm{H'})$$

where: C = fresh clutch weight; H = percentage of clutch hatching; F = number of females; C′ = mean clutch weight; and H′ = annual mean hatch success.

Both the estimate of the breeding success of the population and the total number of chicks hatched were positively correlated with increasing female density. No significant relationship between female numbers and PBS′ means that the area can only support a given productivity, regardless of density. A significant positive relationship indicates that the area can support a higher breeding greenshank density than is often present, and with a gross return on reproductive effort. From the significant reduction in fresh clutch weight with increasing density, the authors hypothesize that competition between females in pre-nesting groups on tidal flats and in the bays of large lochs may have hindered the females' ability to acquire sufficient reserves for egg laying, but the other results suggest that this did not act in a negatively density-dependent manner on the estimate of population productivity.

The factor that seemed to correlate with the density of birds on the breeding ground was the PBS in previous years: when PBS was low, fewer pairs bred in the following 2 years. As the number of birds was also smaller when June of the previous 2 years was cold and wet, the authors infer that the harsh conditions affected chick survival and lowered recruitment to the population over the next 2 years.

These results suggest two things: firstly, if displaced birds are able to establish in another area, increase in population pressure on food resources – especially around lochs and rivers – will promote competition, reduce food availability and intake, and result in the production of less viable eggs. Secondly, however, the study area could support variable densities of birds, and population PBS was positively correlated with increasing density. At no time were all territories occupied; a small but largely constant floating population was present but these birds did not take up vacant territories or those vacated by failed breeders. The authors suggest that these birds might have been prevented from breeding by food supply or competition but no evidence for this is presented. The yearly variations in density were explicable in terms of total population size as a result of breeding success in previous years and neither the density, nor the PBS of the birds was correlated with any factor in the year of breeding, suggesting that the birds did not track proximal conditions closely (no measures of resource availability were made, however).

Many other factors are likely to be involved when considering the effects of forestry in such circumstances but, when combined, these results suggest the possibility that additions to this greenshank population in any given year would have encountered vacant territories, would have had the resources to breed and could have improved the total production output of the local population. Displacement in this hypothetical example, in the previous terminology, would have been under-restrictive to the greenshank of the region.

The results of these case studies suggest that the likely effects of afforestation are complex and depend on local circumstance as well as the ecology of the species concerned. The studies were chosen because the species involved are supposedly particularly at risk from upland afforestation, yet it has been shown that, in some circumstances, population decline with increasing forest area, actually or potentially, may be neither linear nor 'supraproportional' (Thompson *et al.*, 1988) relative to forest area. We have concentrated on potential under-restriction of resources by forestry in the previous sections only because the possibility of this consequence is rarely considered. For instance, in the review of Thompson *et al.* (1988), only linear and supraproportional population decline relative to forest area are illustrated. The theoretical lines of evidence they present to suggest that forestry may be more damaging than its displacive effect would immediately suggest are convincing but the theory requires careful testing. For raven especially, forestry in Wales seems to have resulted in markedly 'subproportional' decline. For greenshank, and other migratory species, which may not suffer density-dependent population regulation on breeding grounds where forestry is taking place (e.g. the Flow Country), subproportional decline is also a theoretical possibility that needs to be considered.

Indirect effects of forestry

The indirect effects of afforestation, edge effects in particular, have probably received more discussion than the direct effects (Avery and Leslie, 1990). Edge effects are of great potential importance: a demonstration of the existence of deleterious effects beyond the forest boundary would effectively mean that we should be assessing the environmental impact of forestry not on the basis of the planted area, but rather planted area plus some additional zone of influence; that is, forestry would be over-restrictive to the disturbed bird community. This restriction could be particularly serious relative to the forest blocks themselves, as it takes a comparatively small zone of addition around the edge to double forest area.

There are various ways in which edge effects may influence bird populations. Forestry may affect the growth of moorland vegetation around the forest boundaries and on open ground within the forests. This may occur directly due to shading, alteration of run-off characteristics, alteration of microclimate, or through competition. Indirectly, forestry often leads to changes in land-use around the forest margins, such as cessation of moor-burning or reduction of grazing pressure, which can alter vegetation characteristics. The edge effect that may have the most profound consequences is an increase in predation pressure in a zone around the forest caused by the influx of forest-adapted predators such as foxes, stoats and carrion crows. Studies on the predation of upland birds are virtually non-existent. Generalizations, therefore, are inadvisable, but it may be a general feature of predative interactions that they are more intense when the prey population exists at higher densities because predators may target such areas (Thompson *et al.*, 1988). This has not, however, been demonstrated around British upland plantations.

Rather than discuss all the possible changes that can occur around forests, we will concentrate on reviewing the research that has tried to assess whether birds actually show any avoidance of forest edges.

EDGE EFFECTS

The first piece of upland research on this problem again stemmed from concern over afforestation in the Flow Country. Stroud and Reed (1986) surveyed six sites in Sutherland, northern Scotland, by walking transects 200 m apart and parallel with the forest edge. Bird observations were recorded on 1:10000 base maps. Vegetation variation was also recorded in 200 m^2 quadrats and the frequency of bird registrations were compared in quadrats at three distances from the forest edge: $<$ 400 m, 400–800 m and $>$ 800 m. An uncontrolled comparison of frequencies in quadrats at different distances from the forest edge showed that for all species of waders combined, breeding occupancy was lower close to the forest than further away. It was found, however, that the eight main vegetation types on the adjacent moorland also differed significantly in their availability with distance from the forest edge, and as waders may select preferentially for different vegetation types, either directly or because the vegetation in this area has different structural characteristics and reflects ground wetness conditions, the result may have been spurious. When this vegetation variation was controlled for, significant edge effects were still observed for dunlin, golden plover, curlew, lapwing and redshank. The results for greenshank were ambiguous. This effect persisted out to 800 m or more from the forest edge. To put this in perspective, a strip of land 800 m wide around a 10 km $\times$ 10 km plantation covers about one-third of the total area of the original block (and plantations in the Flow Country are not square). If such enormous zones of effect were really present around forest blocks, a complete reassessment of the impact of forestry on the waders of this region would be necessary. It would also have serious implications for forestry policy. Large plantations have lower edge/interior ratios, so the results of Stroud and Reed (1986) would suggest that regular, blanket afforestation would be more environmentally sensitive than a similar area in a mosaic of smaller blocks. At the time, this was the opposite of general opinion.

Subsequent to the publication of Stroud and Reed's analysis, certain errors in their analytical procedure were uncovered, and a further bias was reported in Avery (1989). The latter problem concerned the way in which the 200 m quadrats were sampled for the pairwise comparisons between distance zones that formed the basis of the analysis. This resulted in a tendency for the results to be significant in the direction reported.

The research reported in Avery (1989), using similar survey techniques but different methods of analysis showed that, once differences in vegetation characteristics were controlled for, distance from the forest edge itself could account for only very small and non-significant differences in the variations in bird numbers. Probably the most important result from this study concerned the recognition that bird numbers varied between sites adjacent to forests of different ages, before controlling for vegetation effects. The work of Stroud and Reed (1986) showed that waders declined in numbers with distance from the forest edge. Whether this was due to avoidance *per se*, or due to vegetation differences that had been induced by the proximity of the plantation is largely immaterial; the result is less waders on the adjacent moorland.

A third alternative was also possible, however – that the differences in vegetation with increasing age of forest were due to changes in planting strategy over the years. Avery (1989) extracted variables for each of his study sites from maps, reasoning that these could not be influenced by differences in tree age. Three of these variables (highest and lowest contours, and the number of contour lines crossed by a transect line) were shown to be correlated with tree age, but none with distance from the forest edge. This mirrored the variation in vegetation. Thus there were differences between sites associated with the age of the forest development that could not have been a result of the ageing of the trees. Avery concludes that his results do not support the hypothesis that, as a forest ages,

> its cumulative effect on either the birds or the vegetation of the adjacent land increases, but rather...that different types of site have been planted at different times over the past 40 years.
>
> (Avery, 1989)

During the survey that led to these findings, experiments were also performed to assess the likelihood of increased predation of nests with proximity to the plantations (Avery *et al.*, 1989). Chicken eggs were placed at 200 m intervals from the forest edge and visited 1 week later to see if they had been removed. The results mirrored those of the previous work: eggs placed close to the edge of the forest were subject to predation more often than those further away, but this again was closely related to differences in vegetation characteristics. When these differences were allowed for, it turned out that predation was unrelated to proximity. As Avery points out (Avery and Leslie, 1990), this suggests that predation on real nests might be unrelated to proximity to the forest edge, but studying the effects of predation in the real situation is extremely difficult.

Two further studies are of relevance here. Firstly there is the re-analysis of the data from the study of Stroud and Reed (1986), which was published in 1990 (Stroud *et al.*, 1990). Using more appropriate analysis techniques, this study maintained that statistically significant edge effects were apparent for golden plover and dunlin, which were not explicable by variations in the distribution of vegetation types on the adjacent moorland. A significant avoidance of the forest edge was also recorded for all waders combined, although it is difficult to see the justification for treating different wader species as a single group when it is clear from the analysis that most of the species considered do not show edge avoidance. Increasing the sample size of observations in this way may lead to greater statistical significance, but the ecological significance of the result is more difficult to assess. A significant difference in common sandpiper numbers was recorded between distance zones from the forest edge but suggests that numbers are at their lowest at intermediate distances. Also, as the authors point out, with a sample size of three survey sites for this statistical test, the result should be considered with suspicion. Unfortunately, the re-analysis of the data still contains statistical errors (Mark Avery, personal communication). Given the small sample sizes involved, exact probability tables for the non-parametric two-way analysis of variance should have been used (see Siegel, 1956). Of the three results reported as significant in the analysis of bird distributions in relation to distance from the forest edge, disregarding the result for common sandpiper, one (for dunlin) is correct, another (for all waders) is less significant than reported, and the last (for golden plover) is not significant at all. The existence of edge effects in this dataset, therefore, remains questionable.

Another study carried out between 1987 and 1989 (Parr, 1990) presents very similar results to those of Avery (1989). Significantly lower densities of dunlin and golden plover and higher densities of snipe and curlew were recorded next to older plantations (but not adjacent to younger ones) when compared with control areas away from the forest edge. Breeding success was also lower for golden plover and red grouse next to old trees. However, when the data were re-analysed to take into account inherent differences between edge and control areas using map-derived variables in a similar way to the analysis of Avery (1989), it was found that the previously observed differences in density and breeding success were no longer apparent. Also, no significant difference in the number of recorded predator kills was apparent between edge and control areas.

From these studies, the weight of evidence, taking into account methodological considerations in particular, appears to show that edge effects may not be an important consideration in assessing the effect of afforestation on upland birds. Certainly the situation appears less serious than it did subsequent to the publication of Stroud and Reed (1986). More detailed studies are required, however, in a wider variety of geographical locations, before the controversy over edge effects around upland plantations will finally be resolved.

CONCLUSIONS

From this review, it is clear that the effects of afforestation on upland bird populations are not obvious even when considering species that are particularly sensitive. Research is generally unavailable to assess the survival or viability of displaced birds, or the effect on population dynamics. The detail of the response of birds to afforestation, and the relative decline that ensues, is likely to be different from one species to the next, and even within a species, from one area to the next.

Thompson *et al.* (1988) at the end of their review on the population consequences of afforestation, consider the likely effects on the population size of ground-nesting birds on unplanted ground around plantations. The simple graphical models presented give useful and testable 'classical' predictions of the expected effects of afforestation against which the results of empirical studies can be judged. In these models, populations are expected to decline in size due to the interaction of various hypothesized factors operating on displaced birds or the receiver population. These include resource limitation due to reduced territory size, failure to gain a territory through non-reduction of territory size, loss of local experience for displaced birds, increased competition and increased nest and young predation. All of these processes are undoubtedly important in many cases and need to be quantified in each case to assess the full effect of afforestation. As mentioned previously, however, if some of the assumptions about what commonly controls population structure and breeding performance are not realized in a given population or a given area, these processes may not be relevant. The fundamental assumption that underpins the usual methods of predicting the effect of forestry, is that the population size of moorland birds under normal conditions, reflects the carrying capacity of the environment. If, as Evans and Pienkowski (1984) suggest, the carrying capacity of breeding grounds for birds like the threatened waders of Table 9.3 is 'utopian', and that such species do not approach this capacity under normal circumstances, the theoretical basis for such predictions is substantially weakened.

The models of Thompson *et al.* (1988) are specifically tailored to ground-nesting birds and exclude their predatory and scavenging guild, possibly because of the complicated results on raven and buzzard populations in Wales (Newton *et al.*, 1982). Many of the consequences for displaced ground nesters, however, may be similar to those for predatory birds if afforestation is considered in terms of its effects on resources. If area-linked resources are not normally limiting, if the birds do not saturate the available habitat, if range expansion in response to displacement is possible, and if populations are normally regulated outside the breeding grounds, it is feasible that the predicted adverse consequences of afforestation for bird populations may not materialize until a threshold of habitat loss is reached (which may be considerable) and that decline therefore, when considered in terms of the population of a region, may not keep pace with habitat loss. Thus, decline with increasing afforestation may not be either supraproportional or even proportional to forest area. Also, the most straightforward way in which supraproportional decline may come about is through the operation of edge effects. To date, however, no study has shown satisfactorily that edge effects operate to limit the densities or productivity of moorland birds adjacent to upland plantations.

Before one can accurately predict the effect of afforestation on upland bird populations, information is required on the population dynamics of bird populations in forestry affected areas – especially for those species that appear, by reason of their distribution or ecology, to be most at risk. In this way, research can be directed to the best conservation benefit. The most important information required concerns the following:

1. The response of individuals to the displacement 'event' itself: can they re-establish, where and when does this take place, and what are the subsequent reproductive consequences for the individual and the population?
2. The relationship between bird species communities and the availability and use of essential resources. This information is of fundamental importance to the validity of most ecological investigations and the generalizations that such

studies give rise to; it is, nevertheless, the area that has been notoriously neglected by ecologists for much of the history of the subject (Wiens, 1984; 1989).

3. Whether under 'normal' circumstances populations are limited by density-dependent processes due to resource conditions, social behaviour and competitive or predative interactions.

4. The way in which forestry affects resource conditions and inter- and intra-specific interactions, and the ramifications of this in the light of the information on points (**1–3**).

ACKNOWLEDGEMENTS

We are grateful to Mark Avery of the Royal Society for the Protection of Birds for his constructive criticism of the manuscript, and for additional information on edge effects around British upland plantations.

APPENDIX: LATIN NAMES OF BIRD SPECIES

Bunting, Snow: *Plectrophenax nivalis*
Buzzard: *Buteo buteo*
Chough: *Pyrrhocorax pyrrhocorax*
Crow: *Corvus corone*
Cuckoo: *Cuculus canorus*
Curlew: *Numenius arquata*
Dipper: *Cinclus cinclus*
Diver, Black-throated: *Gavia arctica*
  Red-throated: *Gavia stellata*
Dotterel: *Eudromias morinellus*
Dove, Stock: *Columba oenas*
Dunlin: *Calidris alpina*
Eagle, Golden: *Aquila chrysaetos*
Eider: *Somateria mollissima*
Fieldfare: *Turdus pilaris*
Grebe, Slavonian: *Podiceps nigricollus*
Greenshank: *Tringa nebularia*
Goosander: *Mergus merganser*
Goose, Greylag: *Anser anser*
Grouse, Black: *Lyrurus tetrix*
  Red: *Lagopus lagopus*
Gull, Black-headed: *Larus ridibundus*
  Common: *Larus canus*
  Great Black-backed: *Larus marinus*
  Herring: *Larus argentatus*
  Lesser Black-backed: *Larus fuscus*
Jackdaw: *Corcus monedula*
Kestrel: *Falco tinnunculus*
Kite, Red: *Milvus milvus*
Lapwing: *Vanellus vanellus*
Mallard: *Anas platyrhynchos*
Merganser: *Mergus serrator*
Merlin: *Falco columbarius*
Nightjar: *Caprimulgus europaeus*
Osprey: *Pandion haliaetus*
Ousel, Ring: *Turdus torquatus*
Owl, Short-eared: *Asio flammeus*
  Tawny: *Strix aluco*
Oystercatcher: *Himantopus ostralegus*
Peregrine: *Falco peregrinus*
Phalarope, Red-necked: *Phalaropus lobatus*
Pipit, Meadow: *Anthus pratensis*
  Tree: *Anthus trivialis*
Plover, Golden: *Pluvialis apricaria*
  Ringed: *Charadrius hiaticula*
Ptarmigan: *Lagopus mutus*
Raven: *Corvus corax*
Redshank: *Tringa totanus*
Redwing: *Turdus iliacus*
Sandpiper, Common: *Tringa hypoleucos*

  Wood: *Tringa glareola*
Scoter, Common: *Melanitta nigra*
Skua, Arctic: *Stercorarius parasiticus*
  Great: *Stercorarius skua*
Skylark: *Alauda arvensis*
Snipe: *Gallinago gallinago*
Stint, Temminck's: *Calidris temmincki*
Stonechat: *Saxicola torquata*
Teal: *Anas crecca*
Thrush, Mistle: *Turdus viscivorous*
  Song: *Turdus philomelus*
Twite: *Carduelis flavirostris*
Wagtail, Grey: *Motacilla cinerea*
  Pied: *Motacilla alba*
Warbler, Grasshopper: *Locustella naevia*
  Willow: *Phylloscopus trochilus*
Wheatear: *Oenanthe oenanthe*
Whimbrel: *Numenius phaeopus*
Whinchat: *Saxicola rubecula*
Whitethroat: *Sylvia communis*
Wigeon: *Anas penelope*
Wren: *Troglodytes troglodytes*

Other species

Fox: *Vulpes vulpes*
Stoat: *Mustella ermina*

10

A Comparative Study of the Soil Mite Communities (*Acari*) of Wooded and Unwooded Areas in the Basque Country, Northern Spain

M. SALOÑA AND J.C. ITURRONDOBEITIA

SUMMARY

The soils of a variety of different types of habitat have been sampled to investigate the effects of afforestation on the soil fauna. The habitats were chosen to represent different levels of human interference and included natural woodland, non-woodland sites and plantations. Variations in the populations of oribatid soil mites were studied in relation to afforestation and the level of human intervention. The possibility of using oribatid soil mites as indicators of soil condition is investigated.

INTRODUCTION

Edaphic oribatid mites play an important role in the decomposition of organic matter. Decomposition processes tend to be faster if oribatid mites are present, and the soils become more balanced and fertile. As a result, oribatid mites are one of the best biological groups for indicating the ecological condition of the soil, and by implication of the whole ecosystem.

Studies of the effects of abiotic factors on edaphic microarthropods have become more frequent in recent years (e.g. see Lebrun, 1965; Banerjee, 1984; Hagvar, 1984; Wauthy *et al.*, 1989). In addition, there have been important studies of the effects of different forms of agricultural management on the edaphic fauna (Siepel *et al.*, 1989). Valuable information can be derived from these studies about the role of edaphic microarthropods as indicator species, which can be used in the more effective use and management of agricultural resources, and also to assist in the assessment of environmental hazards.

The main aim of the study outlined in this chapter is to provide an introductory assessment of the relationships between the abundance of oribatid mites, soil types and forest management. This will allow us to improve our knowledge of the ecology of oribatid mites and assess their value as bio-indicators.

THE STUDY AREA

The area of study is the Biscay (Vizcaya) region, which is one of the provinces of the Basque country of northern Spain. Over the last 250 years there has been a gradual, although continual, clearance of the woodlands in this region. Deforestation was well documented by the end of the 18th century but the majority of the land surface was still woodland at the beginning of the 19th century, when the first general forest statistics were published. Although these statistics are not accurate, it can be estimated that 84% of the woodland was oak, 13% was chestnut, 2% was evergreen oak and around 1% was beech. The main causes of woodland change in this period were the management of woodland to produce charcoal for iron foundries, the use of timber for ship building and the clearance of woodland and its conversion to agricultural land.

During the 20th century there has been much degradation of the natural landscape and a massive increase in afforestation with coniferous species and eucalyptus in order to improve the amount of home-produced timber. Today about 55% of the surface area of Biscay is woodland. Half of this woodland consists of coniferous plantations and 43% is Monterey pine (*Pinus insignis*). It is clear, therefore, that there have been massive changes in the extent and nature of woodland in the present century. Some problems associated with forest policy and practice in the Basque country are discussed by Groome (1991).

METHODS

Nineteen different soil sites grouped in three study areas were chosen from within the Biscay region (Table 10.1; Fig. 10.1). One of the study areas was on the coast, the second in an agricultural inland valley, and the third was situated in a mountainous area that was partly agricultural and partly covered with semi-natural vegetation. The sites were chosen in order to provide samples of soils from forests with little human intervention, soils from plantations and soils from unwooded areas with different levels of management. The types of soil varied considerably from site to site. In the coastal study area, the eucalyptus forest (2) pine forest (3) and meadow (4) were growing on a clay loam. The holm oak forest (5), mixed deciduous forest (6) and pedunculate oak forest (7) were on a sandy loam, while the dune pine forest (1) was growing on sand. The proportion of clay in the soils increased in the valley and mountain study areas. The Portuguese oak forest (16) pine forest (17) and holm oak forests (19) in these areas had typical clay soils.

Table 10.1. The type of habitat at the sample sites and the sampling dates. (See Fig. 10.1.)

Tree species	Code*	Habitat type	UTM†
Pinus pinaster	DU-1	Dune with pine forest	30TWP048084
Eucalyptus globulus	EU-2	Eucalyptus forest	30TWP094090
Pinus radiata	PI-3	Pine forest	30TWP094090
	PR-4	Meadow	30TWP092088
Quercus ilex	EN-5	Holm oak forest	30TWP092088
Robinia pseudoacacia	BM-6	Mixed deciduous forest	30TWP048033
Quercus robur	RO-7	Pedunculate oak forest	30TWP048033
	PR-8	Meadow	30TVN724851
Quercus robur	RO-9	Pedunculate oak forest	30TVN724851
Pinus radiata	PI-10	Pine forest	30TVN718864
Robinia pseudoacacia	BM-11	Mixed deciduous forest	30TVN762878
Eucalyptus globulus	EU-12	Eucalyptus forest	30TVN780880
Fagus sylvatica	HA-13	Beech forest	30TWN019550
	PR-14	Incipient moorland	30TWN001559
	PR-15	Meadow	30TVN986580
Quercus faginea	RO-16	Portugese oak forest	30TWN001578
Pinus radiata	PI-17	Pine & mixed dec. forest	30TWN001559
	PR-18	Meadow	30TVN998586
Quercus rotundifolia	EN-19	Holm oak forest	30TVN986582

Sample dates for autumn, winter, spring and summer were:
1. Coast 10.10.84, 14.1.84, 23.4.85, 3.7.85;
2. Valley 22.10.84, 25.1.85, 2.5.85, 9.7.85;
3. Mountain 7.11.84, 4.2.85, 16.5.85, 2.8.85.

*Sample site code
†Universal Transversal Mercator System

Soil acidity was greatest in the coastal and valley areas; the lowest pH values being in samples from the eucalyptus forests (2 and 12) and the pine forests (1 and 3). The soils of the mountainous area around Orduna, by way of contrast, had high pH values. Exceptions were the soils of the beech forest (13) and incipient moorland (14), which despite having a mull humus had acid pH values; these were probably due to the high humidity and high content of organic matter. The distribution of humus types was as expected, with the coniferous sites (such as pine forest 3) having more humus, and the deciduous (beech forest 13) a mull-type humus. More detailed results of our soil analysis have been published elsewhere (Iturrondobeitia and Saloña, 1990).

For the analysis of the soil fauna, two samples of 500 cm^3 of soil matter were taken from each of the sampled ecosystems. These samples were taken in the

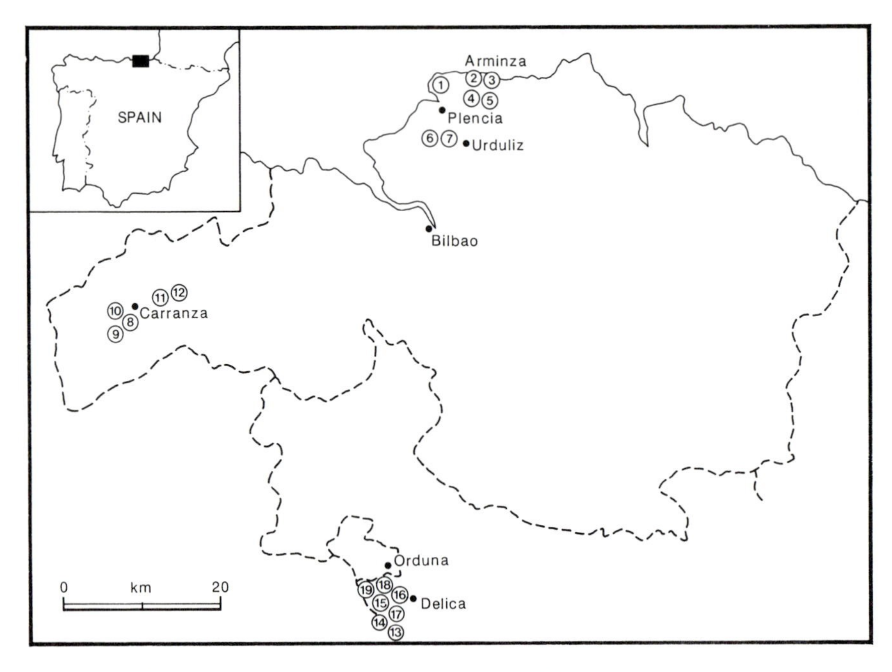

Fig. 10.1. The location of the sample sites. (See Table 10.1.)

autumn of 1984 and again in the winter, spring and summer of 1985. The microarthropods were extracted with the use of a modified Berlese funnel. Several measures of community structure were calculated, including Shannon diversity, evenness, species substitution and Motomura's constant. These were calculated for each community from the species abundance matrix (Table 10.2).

Diversity and evenness were calculated by following the methods described by Cancela Da Fonseca (1969). The phenological species substitution (Iturrondobeitia and Subias, 1981) was calculated as the difference between H_t and $\bar{H}_s$, where H_t is the species diversity of a community calculated using the seasonal samples added together; and $\bar{H}_s$ is the arithmetic mean value of the four different seasonal values. This parameter yields important information about species replacement through time. H_t increases when sample species are different from one season to another while $\bar{H}_s$ remains more or less constant; H_t will decrease if the number of species remains constant season by season because it takes account of numbers of individuals but not numbers of different species. Motomura's constant (Daget, 1979) tells us about the dominance of species in a community and can be interpreted as an indication of the degree of stress or equilibrium. Its calculation has been adjusted by eliminating the tail formed by variations in the number of rare species that can substantially affect the calculated index value. This eliminated tail corresponds to the 5% abundance of less abundant rare species (Ascacibar and Iturrondobeitia, 1984).

Table 10.2. Relationship between the abundance (Ab), number of species (Sp), specific diversity (H_t), phenological species distribution or replacement ($H_t - \bar{H}_s$), evenness (E) and Motormura's constant (KM) of the 19 oribatid communities studied.

Community type	Code	Ab	Sp	H_t	$H_t - \bar{H}_s$	E	KM
Dune and pine forest	DU-01	470	36	4.03	0.77	0.78	0.85
Eucalyptus forest	EU-02	3500	62	4.32	0.44	0.72	0.88
Pine forest	PI-03	3873	65	4.29	0.59	0.71	0.87
Meadow	PR-04	2639	48	3.37	0.46	0.60	0.84
Holm oak forest	EN-05	1337	65	4.72	0.91	0.78	0.91
Mixed deciduous forest	BM-06	694	61	4.91	0.96	0.83	0.92
Pedunculate oak forest	RO-07	1971	73	4.82	0.46	0.78	0.93
Meadow	PR-08	1467	47	4.07	1.08	0.73	0.85
Pedunculate oak forest	RO-09	1493	75	4.64	0.82	0.75	0.89
Pine forest	PI-10	1916	66	4.64	0.58	0.77	0.89
Mixed deciduous forest	BM-11	1097	59	4.78	0.61	0.81	0.91
Eucalyptus forest	EU-12	2397	73	4.61	1.06	0.74	0.90
Beech forest	HA-13	3721	84	4.90	0.64	0.77	0.92
Incipient moorland	PR-14	2127	78	3.86	0.64	0.61	0.88
Meadow	PR-15	1791	58	3.43	0.24	0.59	0.85
Portugese oak forest	RO-16	2067	84	4.23	0.64	0.66	0.89
Pine and mixed deciduous forest	PI-17	2129	69	4.28	0.56	0.70	0.89
Meadow	PR-18	850	31	3.75	0.71	0.76	0.84
Holm oak forest	EN-19	3600	73	4.13	0.52	0.67	0.89

Several statistical techniques have also been employed in order to discover the affinities among communities. These include the use of Horn's index on the logarithmically transformed abundances followed by the classification of those affinities using a form of cluster analysis. The data were logarithmically transformed so that they were all on the same scale (Huhta, 1979) and also 'cleaned' so that only the relevant species were included. Although controversial, this task is necessary. It was undertaken in two ways. First, by eliminating those species that had a high 'ecological niche breadth' (a level of up to 0.8 in a 0–1 scale was chosen); these species do not give discriminative information because they do not exhibit habitat preference. Second, by eliminating the rare species, as had been done in the application of Motomura's constant, by taking out the tail that represented 5% of the total abundance.

Horn's index makes use of a complicated mathematical calculus, which is explained in Brower and Zar (1977). The Lance and Williams (1967) algorithm was used as in the classification of affinities. The result of using all these procedures is a self-explanatory cluster or dendrogram. With the object of verifying the observed dendrograms, Correspondence Factorial Analysis (Daget, 1979; Cuadras, 1981) was also employed; it is a method of multivariate analysis that deals with data ordination in 'n' dimensional space. By means of this analysis, species and sample sites were arranged in a two-way contingency table of observed frequencies and plotted on two axes according to Chi-square distance. The relationships between the plotted elements are indicated by the distances between them.

RESULTS AND DISCUSSION

After separation and identification, the total number of species discovered was 224. There were 39239 individuals with a mean value of 515 mites/l of soil. Table 10.2 shows various measures of community structure for each sample site. The relationship between the total number of individuals in a sample and the number of different species is not straightforward. For example, the coastal pine forest (3) and mountain holm oak forest (19) had the highest number of individuals (3873 and 3600 respectively), but they only had middling diversity values, which were relatively low for woodland communities. The beech forest (13), on the other hand, had one of the highest numbers of individuals together with a high diversity value. The two mixed deciduous forest samples (6 and 11) had relatively low numbers of individuals.

The coastal meadow site (4) was the richest unwooded sample in terms of numbers of individual oribatid mites, but it had the lowest specific diversity. This could be interpreted as an unbalanced community where a few species were dominant. This observation is supported by the low value of Motomura's constant. At the other extreme, the sample from the dunes (1) had the smallest number of individuals but a high diversity; the oribatid community of this sample is closer to those found in meadows than to those characteristic of woodland.

If species richness, another biological factor closely related to species diversity, is considered, it can be seen that the mountain beech forest (13) and Portuguese oak forest (16) are the richest, each containing 84 species. In contrast, only 31 occur in mountain (18) and 48 in coastal meadows (4), and both these have the lowest value of Motomura's constant (0.84).

In general, it can be observed that the 'open' ecosystems such as meadowland and dunes have a poorer oribatid community than the 'closed' woodland ecosystems. In addition, samples from natural woodland sites are richer than samples from plantations. Figure 10.2 shows the values for total and mean diversity, evenness and Motomura's constant in graph form. Sample sites have been placed in order of level of diversity with those having a low diversity on the left. There is a clear link between vegetation cover and the Shannon diversity index. Natural woodland (i.e. beech, oak and mixed deciduous woodland) generally provides a more humid environment and is more diverse than plantations of pine and eucalyptus. These, in turn, are more diverse than meadowland,

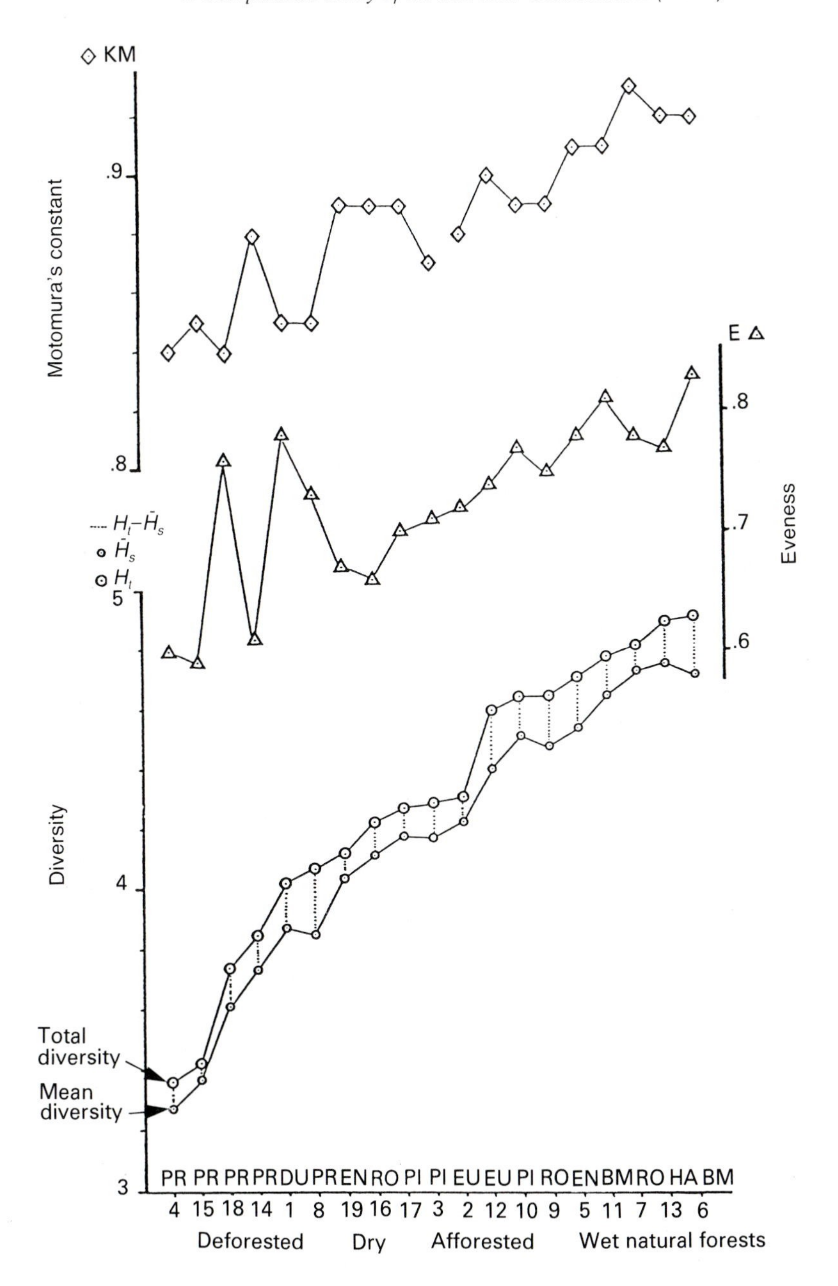

Fig. 10.2. The relationship between specific diversity (H_s), phenological species distribution or replacement ($H_t - \bar{H}_s$), evenness (E) and Motomura's constant (KM) of the nineteen oribatid communities studied. Note that the sites are arranged according to their increasing diversity values.

incipient moorland and dunes. The drier woodlands of Portuguese oak (16) and holm oak (19) are placed between the unwooded sites and the plantations. (Although humidity was not measured directly, it is possible to distinguish qualitatively between wet and dry soils.)

The coastal dunes, despite being planted up with pines, have a richness and diversity equivalent to the meadowland. This can be observed in Figures 10.2 and 10.3. It is the deforested soils, that is the exploited meadowland, which has long been cleared of trees, that provided the poorest samples in this study. These soils were most stressed and had the lowest values of species diversity and

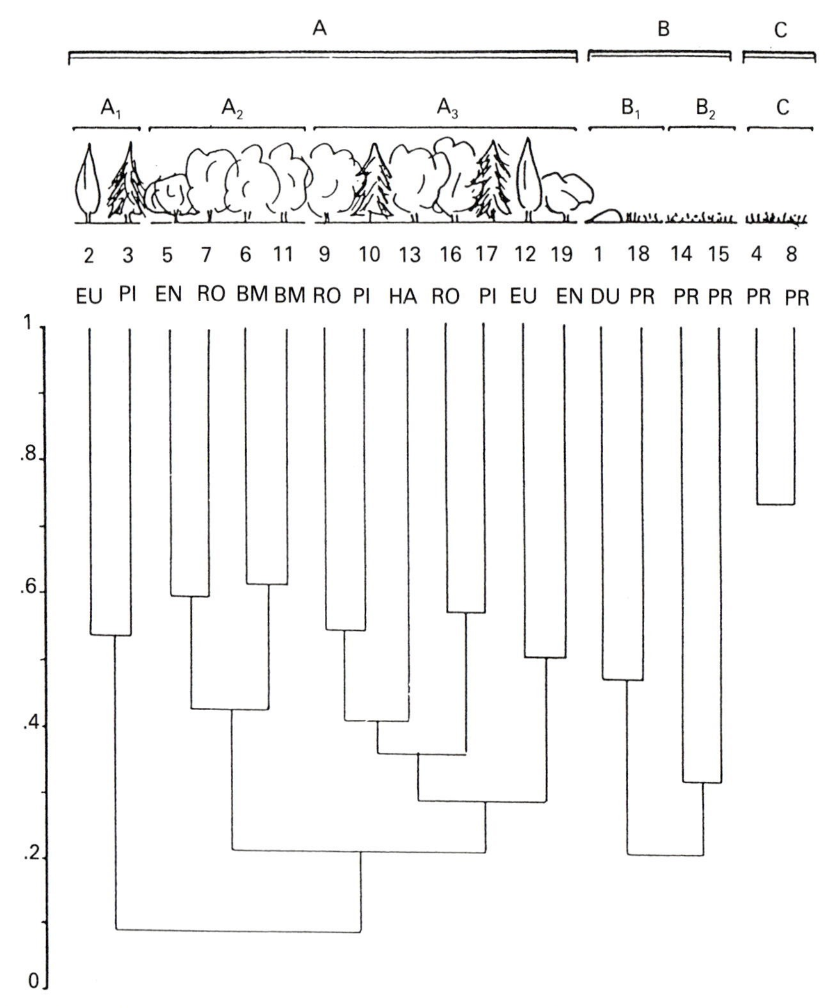

Fig. 10.3. Cluster diagram showing affinities between the oribatid communities. The affinities were calculated using Horn's overlap index and the flexible method of classification was used.

Motomura's constant (see Fig. 10.2). On these sites there has been a decline in biological values following historical deforestation.

There appears at first sight to be no clear link between phenological species substitution (or species replacement through time) and whether the sample originated in woodland or open sites (Fig. 10.2). It appears, however, that species replacement through time is higher in coastal sites where soils are sandy and have a low organic matter content. This indicates that sandy soils have a low stability compared with clay soils.

It was also observed that a well-stratified woodland structure, with a well-developed shrub layer, allows the differentiation of microhabitats, which in turn increases the diversity of microarthropods and provides a greater equilibrium of populations through time. A good example of this is provided by the oak forests (6) and (7) and the beech forest (13) where phenological species replacement is low due to a combination of soil stability and protection given by the shrub layer. In more open areas, such as meadowland and woodland without a shrub layer, high values of seasonal species substitution might be detected. This is not always the case, however, as one of the meadowland samples (15) shows the lowest value for species substitution; a fact that cannot be explained at present.

The clustering and ordination techniques employed to discover the affinities and structure of the 19 oribatid communities resulted in a classification of two main groups (Figs 10.3 and 10.4). One of these was characterized by forest soils and the other by meadow soil communities. The dune sample was distinct from the two groups. It had affinities with the limed meadows as shown in the cluster diagram (see Fig. 10.3) but had its own special character owing to low levels of organic matter, high pH, sandy texture and free drainage.

Within the woodland group a three-fold classification of the oribatid communities can be made based on soil stress or degradation and drainage. First, examples of those woodlands with natural impacted soils (A1) are pine forest (3) and eucalyptus forest (2). These soils are characterized by the absence of a shrub layer, a very acid pH, moder-mor humus type, high conductivity, bad drainage and low Motomura's constants. Many species inhabiting these sites have a high ecological niche breadth – always below 0.79.

The oribatid communities inhabiting soils with little water content (A3) that were affected by farming activity and dry conditions were represented by eucalyptus (12) and holm oak (19). Those inhabiting wet soils (A2) were mainly unaffected by farming activity and examples include mixed deciduous forests and the holm oak wet forest. The richest and most diverse oribatid community was found in the only mountain beech forest ecosystem (13) studied. This is very distinct from the other communities (see Fig. 10.4) and is a good example of a natural and well-preserved ecosystem with a well-balanced soil having its own distinct fauna. It should be pointed out that this site has been the source of several new taxon records.

We suggest that, taking into account all this information, two main factors (among others) influence the distribution and composition of communities of soil

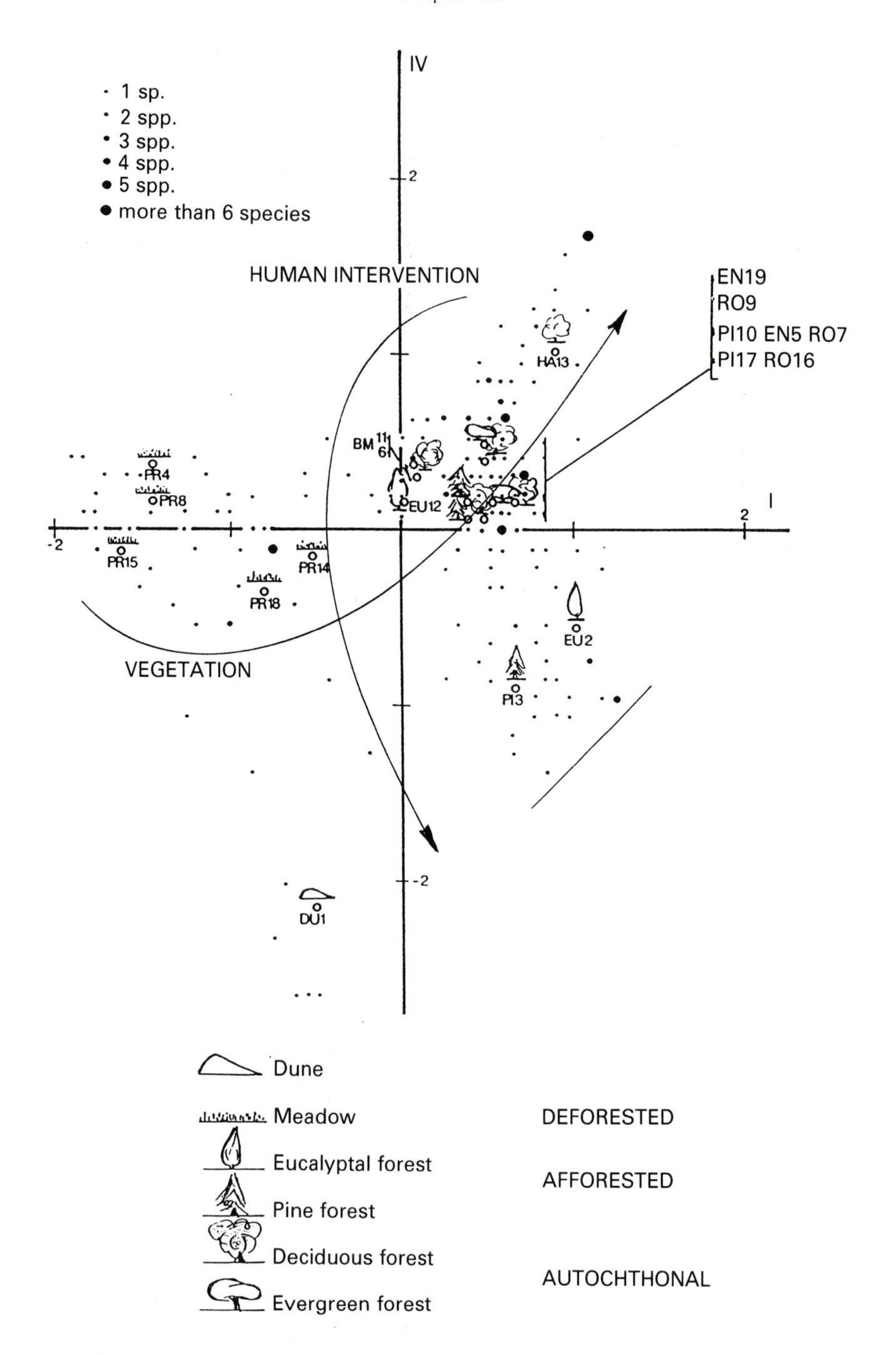

Fig. 10.4. Factorial correspondence analysis in the plane formed by axes I and IV. The arrows show the main observed trends.

oribatids. These are shown diagrammatically by the two arrows in Figure 10.4. The first is the nature of the vegetation. All measures of community structure considered in this study increase as you move from open sites (meadows) towards well-stratified climax woodland ecosystems, such as the beech forest (13). The second main factor is the extent to which the habitat has been affected by human activities. At the lowest level of human intervention (the beech forest, 13) there is a good oribatid community; at the highest level of intervention, such as impacted sites and dunes, there are relatively poor communities. The meadowland communities fall in between these two extremes. The pine (2) and eucalyptus plantations (3) are examples of afforested monoculture ecosystems with trees planted very near each other and without a shrub layer.

Conclusions

Our research has shown that there is a relationship between the nature of woodland and the richness of the soil community. The closer woodland approaches to a 'climax' or natural condition, generally the richer and more diverse is the soil community. Both afforestation, which in recent years in the Basque country has taken the form of monocultures of conifers or eucalyptus, and the conversion of forests to grassland and pastures, result in a decline in the values of measures of community structure of the soil mite community. In comparison, the mite communities of the soils of natural forests are well balanced. Future research should enable us to discover how soil biological factors develop in response to afforestation and other changes in land management. This research should help us to develop measures that can be taken to improve soil conditions.

11

Stream Chemistry and Forest Cover in Ten Small Western Irish Catchments

N. ALLOTT, M. BRENNAN, P. MILLS AND A. EACRETT

SUMMARY

Stream waters in ten small western Irish catchments on granite bedrock were examined for a range of chemical parameters between December 1989 and May 1990. Many of the sites had a higher concentration of hydrogen ions and of inorganic aluminium than is desirable for salmonoid fish. Sites in afforested catchments became the most acidic. Dissolved organic matter, non-marine sulphate and the sea-spray effect all contributed to stream acidity.

INTRODUCTION

County Galway in western Ireland has been famous for the quality of its sea trout and salmon angling since the early 19th century at least (e.g. see O'Gorman, 1845) and angling continues to play an important part in the local economy of the region. In recent times, fish farming of salmon and sea trout has grown rapidly in importance. Forestry is an increasingly important economic activity in County Galway. More than 60 000 ha have already been afforested, mostly with Sitka spruce and lodgepole pine, and EC grant aid is available as an incentive to continued expansion of plantation forestry.

As sea trout and salmon require an abundant supply of running water, it is important to know whether or not plantations impair the quality of surface water.

Evidence from Wales and Scotland indicates that closed-canopy coniferous forests exacerbate acidification effects in surface waters. Streams in afforested catchments have been shown to be more acid and contain higher concentrations of aluminium than streams in unafforested but otherwise similar catchments (Harriman and Morrison, 1982; Stoner and Gee, 1985; Warren, 1989; Bird *et al.*, 1990). The phenomenon has been observed only in 'acid-sensitive' catchments where the bedrock is resistant to weathering and where both bedrock and soils are low in base cations. The principal mechanism seems to be that closed-canopy forest is more efficient at intercepting precipitation (which may contain pollutant acids) than is moorland vegetation.

Forestry mediated acidification of streams has been disputed on the grounds that acid inputs to all catchments have increased since the 1950s (coinciding with the time when many forests were established) as a result of increased burning of sulphur-containing fossil fuels. Acidification in the absence of forestry, to the degree that fish populations have been severely affected, has indeed been documented in extremely acid-sensitive areas such as the south-west of Scotland (Harriman *et al.*, 1987). However, such instances do not invalidate conclusions on forestry mediated acidification so long as the forested catchments are compared with genuinely similar unafforested catchments in terms of geology and soils and other catchment characteristics.

The conclusion that forestry exacerbates the acidification of streams does not rest entirely on comparison of afforested and unafforested catchments. Studies of throughfall chemistry (e.g. see Hornung *et al.*, 1990a) indicate dramatically higher levels of sulphate (SO_4^{2-}) and chloride (Cl^-) in afforested catchments. Elevated levels of mobile anions such as SO_4^{2-} and Cl^- have an acidifying effect because their associated cations (such as ammonia) may be exchanged for hydrogen ions in soils. The acidity of throughfall under spruce up to 25 years old is similar to or lower than that of bulk precipitation (Stevens, 1987) but throughfall and stem flow under 55-year-old spruce is much more acid than that in bulk precipitation (Hornung *et al.*, 1990a).

Chemical changes in precipitation once it reaches the soil are complex and involve neutralization as well as production of acidity. Acidity is reduced by release of aluminium or base cations but can be increased by production of organic acids or exchange of H^+ for Na^+ or Mg^{2+}. The net effect appears to be that soil waters under forests are more acid than their moorland counterparts (Hornung *et al.*, 1990a).

It will be clear so far that forestry mediated acidification is brought about by scavenging of pollutant acids from the atmosphere and therefore should be restricted to areas that are affected by such pollution. The mean annual concentration of non-marine sulphate (assumed to be derived from SO_2 emissions) in the UK declines from more than 75 μEq/l in south-east England to 25 μEq/l in north-west Scotland and the western part of Northern Ireland (Cottrill, 1988). Data are not available for the Republic of Ireland but it is very likely that the annual loading of pollutant sulphur in the west of Ireland is as low if not lower than the western part of Northern Ireland or north-west Scotland.

If the sources of acids that predominate in the UK are largely absent in the west of Ireland, how then might surface waters become acidified in County Galway? There are several possibilities. First, when precipitation that contains a high concentration of sea salt passes through the surface layer of soils (the O horizon) sodium ions can displace H^+ from soil-exchange sites, thereby causing an increase in acidity. The phenomenon where Na^+ of marine origin displaces H^+ from catchment soils and causes an increase in surface water acidity has come to be known as 'the sea-spray effect'. It is to be expected that an area such as County Galway, being on the Atlantic coast and receiving precipitation that originates mainly over the Atlantic, will at times be subjected to these 'salt events'. Second, organic acids are a possible source of acidity in catchments that have extensive peat coverage (Hornung *et al.*, 1990b). Third, while inputs of pollutant acids are probably very low in the west of Ireland generally, occasional easterly winds, bearing acid rain, may cause periodic acid deposition in the area. The coal-fired power station at Moneypoint, which is due south of the study area and emits an estimated 80000 tons of sulphur each year cannot be ruled out as an occasional source of acidity.

It is difficult to say, for several reasons, how much acidity salmonoids can tolerate. First, the different stages of growth are not equally susceptible to acidity. Generally, the early stages of growth require more alkaline conditions – pH values in excess of 4.5 are required during hatching; higher acidity than this deactivates the enzyme chlorionase, which dissolves the egg case and this prevents the hatching alevin from escaping (Crisp, 1989).

Later growth stages are more tolerant of acid conditions. Adults can tolerate a pH of 4.5 or lower if they are acclimatized to such acidity. However, a rapid reduction in pH to below 5.0 (which is very common in softwaters during spate events) upsets the ionic balance. Under these circumstances Na^+ and Cl^- are lost through the gill membrane, which reduces the ionic strength of the blood plasma and, in turn, causes a compensating osmotic flow of water from blood plasma to cells. This causes the blood to become viscous, puts a strain on the circulatory system and causes cardiac overload. Death can occur within hours if the loss of Na^+ and Cl^- exceeds 30% (Wood, 1989). Ca^{2+} in excess of 1 mg/l mitigates the toxicity of H^+ by reducing the permeability of the gills, which helps to prevent loss of Na^+ and Cl^- (McDonald *et al.*, 1980; McDonald, 1983). Unfortunately, catchments that are acidified are also usually low in Ca^{2+} since the increased acidity is often present in the first place due to lack of Ca^{2+} as a neutralizing cation. Mg^{2+} can partly but not completely compensate for a lack of Ca^{2+}.

In Norwegian and other catchments it has been observed that a pH of 5.5 or greater is required to maintain an exploitable and self-sustaining population of salmon (e.g. see Lacroix *et al.*, 1985; Warren, 1989). Choice experiments show that salmonoids will actively avoid waters that have a pH less than 5.5 (Morrison, personal communication), which implies that this pH marks the onset of acid stress. Apparently, salmonoids can tolerate low pH – especially if acclimatized to it, but a pH of 5.5 or greater is highly desirable in salmonoid fisheries. It should also be noted that the minimum permissible pH for salmonoid waters is 6.0 (EC Directive on Freshwater Fish, 78/659/EEC) but derogations are possible under 'exceptional weather or special geographical conditions' (Flanagan, 1988).

The direct effects of acidity on fish is only one way in which acidified surface waters can be degraded as a habitat for salmonoids. Toxic sublethal or even lethal concentrations of labile inorganic aluminium occur commonly in acidified streams. The reason for this is that aluminium ions often replace calcium ions as neutralizing cations in catchments that are low in base cations. It is very difficult to separate the effects of acidity from those of aluminium because high levels of acidity are usually associated with high levels of aluminium.

The chemistry of aluminium in acid waters is complex; in the pH range 4–8 it occurs as Al^{3+}, $Al(OH)^{2+}$, $Al(OH)^{+}{}_{2}$, $Al(OH)^{0}{}_{4}$ or as complexes with fluoride and organic matter; the proportion of aluminium in any of these fractions depends on the concentration of fluoride, dissolved organic matter and the pH of the water. The most important form of aluminium to measure in acidification studies is 'labile monomeric Al' (LM Al) which includes the ionic forms above but excludes aluminium complexed with fluoride or organic matter. Even though LM Al may include several ionic forms, the concentration of LM Al correlates well with the toxicity of this metal in waters (Bull and Hall, 1987). No universal critical threshold of LM Al can be stated because its toxicity depends on the prevailing pH, the concentration of Ca^{2+}, as well as the level of any other heavy metals such as Cu, Pb or Zn that may be present.

Turnpenny *et al.* (1987) concluded, after studying 60 stream populations of brown trout in England (mean pH about 5.0, $Ca^{2+} > 1.0$ mg/l), that levels above 40 μg/l were damaging to trout populations. This is not to say that levels above 40 μg/l are always acutely toxic to salmonoids but levels above this threshold would be undesirable in a salmonoid fishery that has a coincident pH in the region of 5.0. The toxicity of LM Al declines below pH 5.0 (as the toxicity of H^+ increases) until, at pH 4.0, Al helps mitigate the effect of H^+ in a similar manner to Ca^{2+} (Wood, 1989).

The principal purpose of the study presented in this chapter is to ascertain the degree of acidity in surface waters of selected acid-sensitive catchments in County Galway. Attention is paid to the factors that lead to acidity in streams and also to the role, if any, that plantation forestry plays in increasing the acid status of streams. The ten catchments that are reported on here are a subgroup of 25 that have been reported on previously (Allott *et al.*, 1990). The full set spans a large range of geology and size, whereas the subgroup here occurs on granite only and are all less than 15 km^2. Site locations are given in Fig. 11.1. The catchments extend up to 15 km inland from the coast and comprise low hills (up to 327 m) and plateaus. They are noted salmon and sea-trout fisheries. The area experiences a temperate oceanic climate that is somewhat exaggerated compared with most of Ireland because of its close proximity to the Atlantic Ocean. Mean annual rainfall is about 1200 mm and the mean annual number of rainy days about 175 (*Atlas of Ireland* – 1978). Winds are generally from the south-west or west. The mean daily temperature is about 6°C in January and 15°C in July. The dominant soil type is peat. Table 11.1 gives the principal characteristics of the catchments.

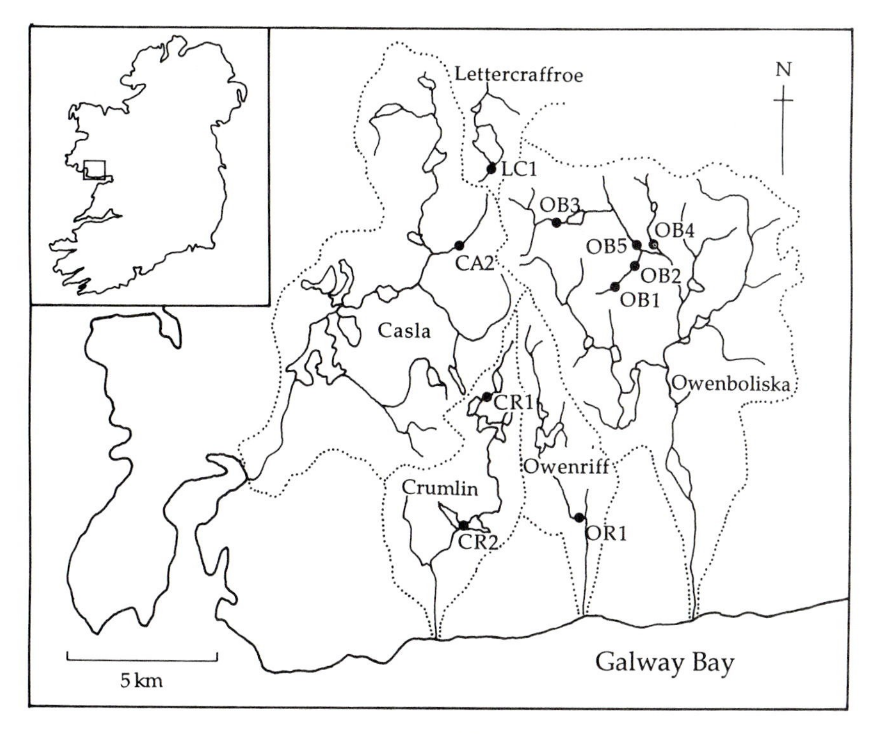

Fig. 11.1. Location of catchments and sampling sites on the Galway granites, western Ireland.

Table 11.1. Locations and basic characteristics of the site catchments in this study. Catchment characteristics (area, percentage forest cover) were computed using a Geographical Information System. The percentage forest cover refers to closed-canopy plantation forest – information that was gleaned from LandSat images.

Site code	Catchment	Grid reference	Area (km^2)	Percentage forest
CA2	Casla	1045 2342	2.4	2
CR1	Crumlin	1056 2292	1.9	0
CR2	Crumlin	1048 2250	11.9	0
LC1	Lettercraffroe	1056 2367	0.3	54
OB1	Owenboliska	1099 2329	0.5	96
OB2	Owenboliska	1107 2336	1.6	93
OB3	Owenboliska	1081 2350	6.6	17
OB4	Owenboliska	1113 2346	3.6	0
OB5	Owenboliska	1109 2342	14.6	29
OR1	Owenriff	1086 2253	12.4	1

METHODS

Water chemistry

The sites were sampled at approximately 2-weekly intervals between December 1989 and May 1990 and samples were analysed for a range of chemical parameters. Samples were taken at sites by immersing and filling completely a 1 l plastic bottle. The high-density polyethylene bottles had been pre-soaked in distilled water prior to being used for the first time and were pre-rinsed with water from the site prior to being filled with the sample. Bottles were allocated to sites for the study period (to prevent the possibility of cross-contamination between sites) and were rinsed with distilled water and stored in the dark between sampling trips. A more rigorous cleaning procedure could not be employed without the risk of contaminating the bottles with one or more of the ions being measured. Samples were transported in cooler boxes. Water temperature and water level were noted. Water level was recorded as being very low (drought conditions), low, medium, high or very high (spate conditions). Duplicate samples were taken at three sites on almost all sampling occasions as a quality control exercise.

Urgent analytical work was carried out as soon as possible (within a few hours) on return to the field laboratory, which was established temporarily in the lodge of Fermoyle House, County Galway. This included determination of pH and alkalinity, as well as filtering (using a Whatman GF/C) and subsampling for further analysis in the Dublin laboratory. The fractionation, by ion exchange, of the samples for later aluminium analysis was also carried out at this point. Subsamples for calcium, magnesium, sodium and potassium were stored in 50 ml polyethylene bottles that had been pre-soaked in 10% hydrochloric acid and rinsed with distilled water. Subsamples for total organic carbon were stored in similar bottles, which had been pre-soaked with 20% nitric acid; 0.045 ml of 50% nitric acid per 50 ml sample was added as a preservative on return to Dublin. A third 50 ml subsample was taken in 50 ml polyethylene bottles, which had been soaked in distilled water; this was used for the determination of nitrate, ammonia, chloride and sulphate.

A glass electrode (Russell CTL/LCW) that has been designed to work well in low ionic strength solutions was used to determine pH. A standard electrode had to be used in the second March sampling as a result of accidental breakage of the Russell CTL/LCW, which could not be replaced in time for the sampling (we consider the pH and alkalinity results for this period to be less reliable than at other times). The electrode was calibrated with low ionic strength buffers (from Reagcon Ltd), which are referenced to buffers from the National Bureau of Standards, Washington DC. The performance of each of the Russell electrodes was determined according to Davison (1990). A zero stirring shift (i.e. obtaining the same reading whether the sample is stirred or quiescent), rapid response and returning of near-correct values (within 0.01 pH units) on dilute acids indicated that the Russell CTL/LCW was suitable for low ionic strength work. Standard electrode characteristics such as the slope and isopotential point were also near ideal. Samples were stirred slowly using a magnetic stir bar until a stable reading was obtained.

Alkalinity was determined by Gran titration (Mackereth *et al.*, 1978) on 100 ml subsamples in a narrow-necked glass bottle. This arrangement minimized the headspace over the sample and therefore reduced exchange of carbon dioxide between sample and atmosphere during the titration. The samples were titrated to three end-points between pH 4.3 and 3.8. The titrant was 0.02 N sulphuric acid, which was dispensed from a mechanical burette (Metrohm E485) into the sample (under the liquid surface) via a teflon tube. Calculations were made according to Mackereth *et al.* (1978) using a simple computer program (Turbo Pascal).

Chloride and sulphate were determined by ion chromatography using a Dionex 2000 series system with the AS4a separator column. Calcium, magnesium, sodium and potassium were analysed by atomic absorption after adding 4 ml of 10% lanthanum per 100 ml sample to control for ionization and interference. Dissolved total organic carbon was determined using a dedicated carbon analyser (Dohrmann).

A 3.5 ml subsample of filtered water (using a Whatman GF/C), which was pipetted into a polystyrene tube, was used for total monomeric aluminium determination. A second aliquot of filtered water was passed through an ion-exchange column (Amberlite IR 120, Na^+ form and 14–52 mesh mixed with 1% of the corresponding H^+ form) to separate the labile monomeric fraction from the non-labile one (procedure adapted from Driscoll (1980) by the Department of Agriculture and Fisheries, Pitlochry, Scotland). Both fractions were preserved with 4 M sulphuric acid (0.035 ml per 3.5 ml sample) on return to the Dublin laboratory. Total monomeric aluminium was determined on the two fractions using an FIA adaption (Tecator Method ASN 78-01/85) of the catechol violet method (Dougan and Wilson, 1974). The result from the filtered subsample is referred to as total monomeric aluminium (i.e. the portion that is easily unbound and made reactive as opposed to being unavailable in, for example, clay minerals).

The portion that passes through the ion-exchange resin is referred to as the non-labile monomeric portion. This fraction is either bound to organic matter or complexed to ions such as fluoride. The bulk of this portion is probably bound to organic matter in western Irish waters from peaty catchments and hence is frequently referred to as 'organic' aluminium (OM Al). The ionic, reactive portion (which is retained on the resin) is the difference between the other two fractions and is referred to as labile monomeric aluminium (LM Al). It is the LM Al fraction that is of most concern in acidification studies since it is this form that is toxic to organisms.

RESULTS

Anticyclonic weather with light easterly winds predominated in November and early December 1989. Weather during January, February and early March 1990 was dominated by a series of cyclones that originated over the Atlantic and caused several episodes of stormy wet conditions. The water level in streams was high on each sampling occasion during this period – particularly so at the beginning of February. The mean concentration of H^+ in rainfall (January to March 1990) at Thallabaun, County Mayo (50 km north-west of the study area), was 1.1 μEq/l (pH 5.95) (Farrell, personal communication); the catchments are therefore very unlikely to have received inputs of acid precipitation during this period.

Levels of Cl^- were very high (Table 11.2) compared with those reported from western Wales (Edwards *et al.*, 1990), western Scotland (Harriman and Morrison, 1982; Harriman *et al.*, 1987) and south-west Norway (Skartveit, 1981). Cl^- values were especially high between mid-January and mid-March; it is probable that rainfall over this period, being derived from a series of deep Atlantic depressions, was highly charged with sea salts since there is no other likely source of Cl^-.

Sites with the highest percentage forest cover had the largest concentrations of Cl^-, which suggests that the coniferous forest intercepted marine-derived atmospheric sea salts more efficiently than peatland vegetation. This is not surprising since it entails the same process as the interception of atmospheric pollutants by coniferous forests.

Comparison between the total and non-marine values of Ca^{2+} and Mg^{2+} (see Table 11.2) shows that essentially all of the Mg^{2+} was from marine sources, whereas up to half of the Ca^{2+} was derived from the catchments. As a result of the pronounced marine influence, the high levels of hardness (Ca^{2+} and Mg^{2+}) did not give rise to proportionately high alkalinities as would normally be the case in inland waters. The alkalinity of these waters seems to be almost entirely dependent on catchment-derived Ca^{2+}. Based on this, sites in the Owenboliska catchment are less vulnerable to acidification than sites in any of the other catchments.

The highest levels of H^+ were reached in the sites with the greatest upstream cover by coniferous forest. This was despite the fact that these sites (OB1 and OB2) are among the least susceptible to acidification in the series. Site LC1 became almost as acidic as site OB2, although it has a lower upstream forest cover; this is explained by the fact that LC1 is much more susceptible to acidification (mean NM Ca^+ of 30 μEq/l) than OB2 (mean NM Ca^{2+} of 91 μEq/l).

Trends in important chemical parameters for two representative sites, one afforested and one unafforested, are given in Figure 11.2. An increase in H^+ occurred at OB1 between early- and mid-December 1989. This increase was associated with increases in TOC (17.9–21.3 mg/l) and NM SO_4^{2-} either of which could have caused the rise in H^+. The NM SO_4^{2-} could have been

Table 11.2. Summary of the chemical characteristics of the study sites: mean with range underneath. All units are μEq/l except for total organic carbon (TOC) which is given in mg/l and labile monomeric aluminium (LM Al) in μg/l. Positive non-marine (NM) values (e.g. NM Ca^{2+}) indicate a source of the ion in the catchment, while negative NM values (e.g. NM Na^{+}) indicate that the sea-salt ion is retained on catchment soils.

	H^+	Alkalinity	Cl^-	Ca^{2+}	Mg^{2+}	NM Ca^{2+}	NM Mg^{2+}	NM Na^+	NM SO_4^{2-}	TOC	LM Al
CA2	7.4	28	876	101	175	63	2	−59	4	4.4	17
	(0.3-23.2)	(−24-142)	(412-1478)	(55-150)	(91-304)	(37-94)	(−16-14)	(−139-32)	(−15-48)	(3.0-8.4)	(0-63)
CR1	15.2	−8	915	88	198	50	11	−59	19	4.5	22
	(5.7-22.7)	(−26-8)	(463-1145)	(50-120)	(107-255)	(32-68)	(−5-19)	(−133-37)	(−46-61)	(2.8-8.4)	(7-41)
CR2	13.3	−7	870	72	181	36	2	−56	18	4.6	20
	(6.7-22.7)	(−23-37)	(468-1204)	(45-100)	(99-255)	(27-51)	(−8-11)	(−142-28)	(−50-55)	(2.8–8.0)	(0-40)
LC1	51.2	−62	1128	74	212	30	−6	−83	7	7.2	43
	(11.9-82.5)	(−108-0)	(415-1977)	(35-135)	(91-428)	(11-60)	(−32-37)	(−233-68)	(−24-82)	(4.4-12.2)	(25-83)
OB1	72.4	−83	1226	120	236	70	−3	−71	12	13.3	73
	(8.4-119.3)	(−144-12)	(558-1966)	(55-185)	(123-395)	(30-110)	(−37-30)	(−263-85)	(−63-118)	(7.3-22.4)	(30-114)
OB2	50.7	14	1371	172	245	91	−18	−132	−15	8.9	86
	(0.6-90.5)	(−108-500)	(1072-1780)	(90-374)	(181-329)	(48-209)	(−25-5)	(−265-−34)	(−52-6)	(6.5-11.2)	(55-116)
OB3	21.4	−8	955	113	195	68	7	−90	5	6	63
	(0.3-45.4)	(−138-178)	(386-1360)	(55-200)	(99-288)	(37-119)	(−4-23)	(−171-58)	(−64-66)	(3.2-12.3)	(44-91)
OB4	2.1	53	756	112	153	72	−2	−54	1	3.5	18
	(0.3-6.6)	(−6-278)	(265-1388)	(55-240)	(66-280)	(41-131)	(−15-14)	(−157-39)	(−26-52)	(1.7-5.8)	(5-31)
OB5	16.5	8	923	108	195	66	10	−74	18	6.5	44
	(0.3-42.3)	(−48-128)	(381-1326)	(55-180)	(99-272)	(37-105)	(0-24)	(−179-66)	(−15-84)	(3.8-12.1)	(12-62)
OR1	5.2	13	928	101	190	62	3	−66	10	4.8	20
	(0.5-12.9)	(−14-54)	(468-1258)	(60-145)	(99-255)	(42-75)	(−17-10)	(−151-23)	(−60-57)	(2.7-8.0)	(0-40)

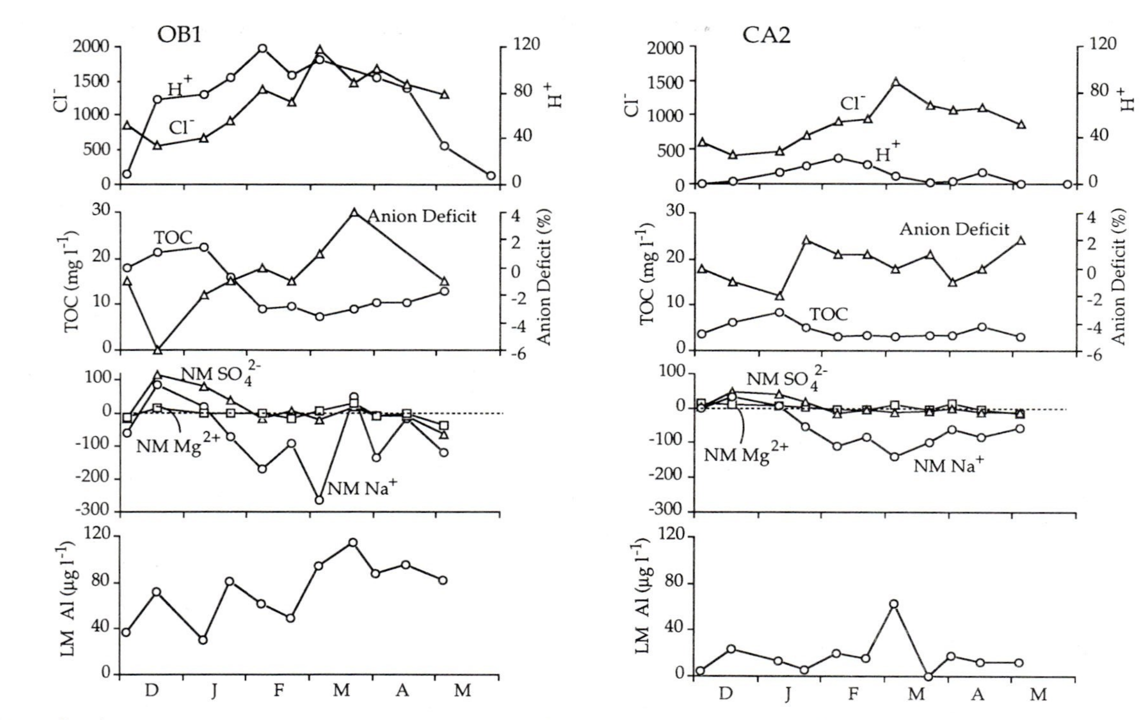

Fig. 11.2. Details of important chemical parameters at two of the study sites. Units are μEq/l unless otherwise stated. The catchments of both sites are granite with peat cover. The catchment of site OB1 is heavily afforested (96% closed-canopy coniferous) while that of site CA2 is almost unafforested. The peaks in H^+ at site OB1 coincide with high concentrations of Cl^- (indicating a large input of sea salts to the catchment) and with large deficits of Na^+ which implies the sea-salt effect as a source of H^+. The effect is discernible but much less extreme at unafforested site CA2.

deposited as dry deposition in the weeks prior to the first sampling or could have been produced by oxidation of soil sulphur – weather conditions in the weeks prior to the first sampling were conducive to both processes. However, the large deficit of anions implies that organic acids were important.

The highest concentrations of H^+ occurred at the beginning of February and March by which time TOC had declined to the lowest measured at OB1, implying that organic acids became less important as a source of H^+. This assertion is supported by the fact that the anion deficit had returned to near zero by the beginning of February. Excess SO_4^{2-} is also an unlikely source of H^+ since its concentration had also returned to zero. However, it can be observed that the maxima in H^+ coincide with minima in NM Na^+ and high Cl^- values, which implies the sea-salt effect as a source of H^+. The negative values of NM Na^+ imply that Na^+ is retained in the catchment when H^+ is released. The effect was smaller at the beginning of March compared with the beginning of February, perhaps because of the reduced soil reservoir of H^+ by then. Na^+ was the principal exchanging cation; Mg^{2+}, the second most abundant cation in sea-water, was not exchanged to a noticeable extent (see Fig. 11.2). The sea-salt effect has been observed to a lesser extent in south-west Scotland (Langan, 1989) and details of it are explained, in the context of its ameliorating effect on soil acidity, by Wiklander (1975).

The trends in chemical parameters were similar at all sites but the degree of change was not so dramatic at the less acidic sites; a comparison between CA2 and OB1 (see Fig. 11.2) illustrates this point. The acidified sites in this study at times displayed H^+ and LM Al levels that are outside desirable limits for salmonoid fisheries. Worse still, peak levels of acidity were recorded in February–March when salmonoid eggs are likely to hatch; $> 32\ \mu$Eq/l of H^+ (pH 4.5) is lethal to hatching salmonoid eggs (Crisp, 1989). LM Al levels were appreciably lower than those found in acidified Welsh and Scottish sites, probably because western Irish soils are known to be low in aluminium (Jeffrey, 1987).

Discussion

The principal importance of our results is that they demonstrate how an undesirable degree of acidification can occur in susceptible catchments that were hitherto thought immune from acidification because of their remoteness from sources of air pollution. What is not clear is the frequency with which salt-bearing Atlantic storms promote undesirable levels of acidity in western Irish streams and, clearly, a longer-term study is needed to establish this. Over a long period, stream acidity is likely to be caused by a combination of the following factors:

- sea-salt episodes;
- organic acidity arising from the oxidation of peats;

- dry deposition of sulphate; and
- acid rain.

Salt-induced acid episodes, such as described in this study, have been identified in Scotland (Harriman and Wells, 1985; Langan, 1989) and the effect has been demonstrated experimentally in Scandinavia by irrigating a small catchment with dilute sea water (Wright *et al.*, 1988). Western Irish catchments, being so far west, must rank among the most susceptible in Europe to salt-driven acid episodes. However, salt-driven acidity in streams differs from acidity due to acid rain because it is completely reversible. When the concentration of salt is high in percolating soil water, sodium ions displace hydrogen ions, which enter solution as hydrochloric acid. However, if the concentration of sea salt is subsequently lowered then the process will be reversed and the concentration of hydrogen ions in drainage water will become less. The effect of sea salts is therefore to increase stream acidity during salt events and to lower stream acidity at other times. By contrast, inputs of pollutant acidity are essentially unidirectional and always bring about increased acidity in catchment soils and waters unless sufficient buffering is produced by weathering of minerals.

Acid episodes arising from inputs of pollutant sulphate to catchments will arise in contrasting weather conditions. First, an anticyclone centred over continental Europe causes an easterly airflow over Ireland; these circumstances may be expected to lead to deposition of pollutant sulphate in dry form throughout Ireland. There is good evidence that much of this deposition occurs in the form of ammonium sulphate, which is acidifying because of exchange with hydrogen in soils. The evidence in this study is that such an episode occurred in the initial sampling period. Second, inputs of acid rain may be expected in the west of Ireland when rain is derived from cyclones whose centres pass to the north of the country. No such conditions occurred during the course of this study but they do occur from time to time.

It is unfortunate that there is no direct method of determining the contribution of organic acids to stream water acidity. However, the coincidence between a high level of dissolved organic carbon and a deficit of anions (presumably due to the unmeasured organic acid anions) represents strong circumstantial evidence for the presence of organic acidity. If the concentration of dissolved organic carbon can be taken to be an index of the concentration of organic acids then it is clear from this study that they are most important during high water conditions following a period of dry weather.

This study has found an association between high levels of stream acidity and high levels of forest cover in catchments. The association, which is based on results from a small number of sites, does not in itself prove a causal link between forestry and stream acidity. However, a causal link is strongly implied when one considers the scavenging phenomenon that is known to occur in mature forests. The increased leaf surface area of a mature forest will tend to intercept any atmospheric aerosol, whether it contains acid rain or acidifying ions such as sea salts or particulate sulphate. Furthermore, the practice of draining peatland soils prior to establishment of trees will promote oxidation of the peat and lead to higher levels of organic acidity. It is therefore not surprising that the most heavily afforested catchments in this study had the most acid streams. Further work is clearly required to substantiate or refute the link between forestry and stream acidity in western Ireland. If plantation forestry increases the acidity of streams, it is clearly ill-advised to promote afforestation of vulnerable catchments.

ACKNOWLEDGEMENTS

We gratefully acknowledge funding from Central Fisheries Board, Western Regional Fisheries Board, Irish Forest Service, Coilte Teoranta, Irish Salmon Growers' Association, Western Game Fisheries Association and Smurfit Natural Resources Ltd. We also thank William Dick of duQuesne Ltd who coordinated the funding.

12

Ecological Changes Following Afforestation with Different Tree Species on a Sandy Loam Soil in Flanders, Belgium

B. Muys and N. Lust

Summary

The ecological effects of afforestation have been evaluated by comparing five 20-year-old homogeneous forest stands, planted on former meadowland, with the adjacent meadow and two nearby 70-year-old forest stands. About 50 physical, chemical and biological variables were measured and treated using multivariate ordination and classification techniques.

The decisive factor explaining differences of earthworm activity, litter breakdown and soil nutrient status between stands was the litter quality of the tree species. The results allowed us to predict the humus profile succession after afforestation of meadows on fairly rich substrates. The silvicultural and forest policy implications of these findings are discussed.

Introduction

An afforestation policy should outline what goals and restrictions have to be established when creating a new forest on former agricultural land. In the case of afforestation within the framework of the European set-aside policy, three important interest groups are likely to have rather different requirements. Farmers and agriculturalists are likely to require that the areas afforested should be easily convertible back to agriculture – they want soil fertility to be conserved. Conservationists are concerned about the effect of afforestation on the species diversity of sites. Modern forestry on a bio-ecological basis will strive after forest stability; as this is a necessity for multifunctional forest use and sustained yield.

The silvicultural techniques used to realize these different objectives should be based on a thorough knowledge of the ecological effects. Particular attention should be paid to the choice of tree species as this has a crucial influence on soil development (Wittich, 1948; Rennie, 1962; Miles, 1981a; 1981b).

It is known that a certain accumulation of litter takes place after afforestation with tree species having a leaf litter that takes a relatively long time to decompose (Ovington, 1956; Babel and Benckiser, 1975; Yeates, 1988). The intensity and reversibility of the soil degradation brought about by afforestation is still a matter of discussion. Based on the observation of young conifer plantations, Ovington (1953) considered that the additional expected soil deterioration was serious enough to suggest that tree species should be chosen on the basis of their likely effect on soil conservation rather than simply to maximize wood production. Long-term trends derived from these early stages of succession could, however, be very misleading. This is because, as a consequence of successive thinnings and the final clearcut, the accumulated litter decomposes and liberates the immobilized minerals. This process can lead to the complete recovery of the soil at the end of the rotation (Page, 1962; Miller, 1988).

Tree species with an easily decomposed leaf litter, on the other hand, are believed to be able to reverse soil degradation by their ability to carry up leached base cations (Brückner *et al.*, 1987) or to increase mineral weathering rates (Miles, 1981a). Rehfüss *et al.* (1990) noticed a gradual biological regeneration of the soil after afforestation of farmland with poplars and willows. Other research, however, minimizes the soil-improving properties of certain broadleaved species (Petch, 1965).

This mini-review of the literature seems to indicate that the ecological effects of afforestation are dependent not only on the tree species but also on the initial soil conditions. Our study will try to obtain a better understanding of such ecological effects following the afforestation of a fairly rich meadow on sandy loam soil; most other studies have dealt with marginal lands.

METHODS

Five homogeneous forest stands planted almost 20 years ago on meadowland were compared with a meadow and two old forest stands nearby (Table 12.1). The comparison included the physical and chemical properties of the soil, the earthworm communities and the production and decomposition of the litter material (Table 12.2). A detailed description of the site and methods used was given in the publication by Muys *et al.* (1992). A global principal components analysis (PCA) was used in which the number of variables was reduced by pre-treating the chemical analyses of soil, holorganic layer, leaf and herbal litter in four subPCAs from which the two main principal components were extracted.

Table 12.1. Listing of the investigated stands.

Dominant vegetation	Year of origin	Former land-use	Symbol
Recent plantations			
1. *Quercus palustris*	1970	Meadow	QPA
2. *Tilia platyphyllos*	1970	Meadow	TIL
3. *Prunus avium*	1970	Meadow	PRU
4. *Alnus glutinosa*	1970	Meadow	ALN
5. *Fraxinus excelsior*	1970	Meadow	FR1
Old forest stands			
6. *Quercus robur*	1920 (approx.)	Forest	QRO
7. *Fraxinus excelsior*	1920 (approx.)	Forest	FR2
8. Meadow	?	Meadow	MEA

Source: Muys *et al.*, 1992.

Table 12.2. Listing of the evaluated variables.

Variable	Symbol	Variable	Symbol
Annual litterfall	LTOT	Soil clay content	CLAY
Annual herb production	HTOT	Soil loam content	LOAM
Herbal ground cover	HC	Soil sand content	SAND
Epigeic earthworm biomass	EPI	Soil drainage class	DRAIN
Endogeic earthworm biomass	ENDO	Vegetation humidity index	MF
Anecic earthworm biomass	ANEC	Soil chemistry pca axe 1	SAXE1
Total earthworm biomass	TPRO	Soil chemistry pca axe 2	SAXE2
Microarthropod density	ARTHR	Holorganic layer chemistry pca axe 1	OAXE1
Litterbag decomposition	DECO	Holorganic layer chemistry pca axe 2	OAXE2
Deco coefficient of Jenny	JENNY	Leaf litter pca axe 1	LAXE1
Macrofauna activity index 1	MAI1	Leaf litter pca axe 2	LAXE2
Macrofauna activity index 2	MAI2	Herbal litter pca axe 1	HAXE1
		Herbal litter pca axe 2	HAXE2

Source: Muys *et al.*, 1992.

RESULTS

The global PCA detected the biological and chemical soil fertility as the most important source of variation between the stands (42%) on the first axis. The variables mainly contributing to the formation of this axis were litter decomposition (DECO, JENNY), earthworm biomass and activity (TPRO, ENDO, MAI1) and nutrient reserves in litter, holorganic layer and soil (LAXE1, OAXE1 and SAXE1) (Fig. 12.1a).

The independent variables *in casu* soil humidity (DRAIN, MF) and texture (CLAY, SAND, and LOAM), intervene little or not on the first axis, which suggests that the tree species is the main cause of variation. The classification of the stands regroups both the young and the old *Quercus* stands on the poor side of

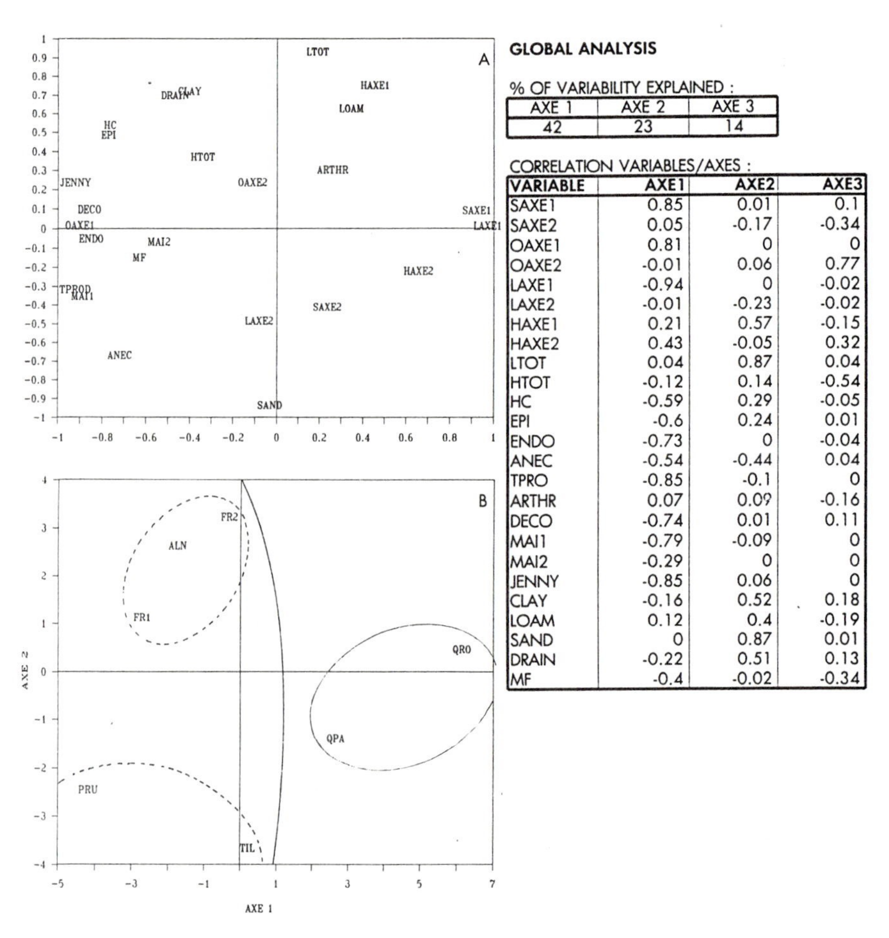

GLOBAL ANALYSIS

% OF VARIABILITY EXPLAINED :

AXE 1	AXE 2	AXE 3
42	23	14

CORRELATION VARIABLES/AXES :

VARIABLE	AXE1	AXE2	AXE3
SAXE1	0.85	0.01	0.1
SAXE2	0.05	-0.17	-0.34
OAXE1	0.81	0	0
OAXE2	-0.01	0.06	0.77
LAXE1	-0.94	0	-0.02
LAXE2	-0.01	-0.23	-0.02
HAXE1	0.21	0.57	-0.15
HAXE2	0.43	-0.05	0.32
LTOT	0.04	0.87	0.04
HTOT	-0.12	0.14	-0.54
HC	-0.59	0.29	-0.05
EPI	-0.6	0.24	0.01
ENDO	-0.73	0	-0.04
ANEC	-0.54	-0.44	0.04
TPRO	-0.85	-0.1	0
ARTHR	0.07	0.09	-0.16
DECO	-0.74	0.01	0.11
MAI1	-0.79	-0.09	0
MAI2	-0.29	0	0
JENNY	-0.85	0.06	0
CLAY	-0.16	0.52	0.18
LOAM	0.12	0.4	-0.19
SAND	0	0.87	0.01
DRAIN	-0.22	0.51	0.13
MF	-0.4	-0.02	-0.34

Fig. 12.1. Global Principal Components Analysis (Muys *et al.*, 1992). A, ordination of the variables. B, ordination/classification of the stands.

the first axis; the other stands, together with the meadow on the rich side (Fig. 12.1b). The origin of the differences observed was found in the classification of leaf litter, based on the ordination of leaf litter quality. The quality of the *Quercus palustris* leaf litter is low and very similar to that of *Quercus robur* (Fig. 12.2).

The litter quality has its repercussions on the decomposition rate. An ANOVA of the holorganic layer of the five young plantations (Table 12.3) found significantly ($P < 0.001$) higher amounts of litter under *Quercus palustris* (727 g/m^2) than under the other species (5–96 g/m^2), except *Tilia platyphyllos* (136 g/m^2), where a certain accumulation took place caused by *Fagus* leaves blown in from the adjacent old stand.

The organic acids produced in the accumulating fermentation layer migrate into the mineral soil and are expected to affect its nutrient status. Although the stand classification (based on the ordination of their soil nutrient concentration) still clusters the *Quercus palustris* stand together with the other young stands and the meadow at the rich side of the first axis (Fig. 12.3), its pH value (at 0–5 cm) is already significantly ($P < 0.001$) lower (pH 4.97) than that of the young *Fraxinus* (pH 6.08), *Prunus* (pH 5.54) and *Alnus* (pH 5.51) stands. It is also somewhat, but not significantly, lower than that of *Tilia* (pH 5.26) and the meadow (pH 5.18) (Table 12.4). It is interesting that all the young stands and the meadow too had a significantly lower pH than the original meadow (pH 6.4) (De Coninck, 1972). This overall pH decrease is mainly due to the lack of artificial fertilization following the planting of the trees.

A decreasing pH has a negative influence on the earthworm biomass ($r = 0.72$, $P < 0.001$). The ANOVA of the earthworm biomass ($P < 0.001$) detected significantly lower values under *Quercus palustris* (344 g/m^2) than under

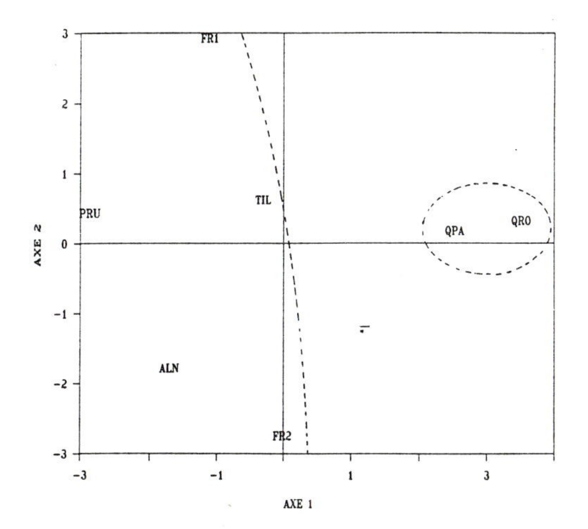

CHEMISTRY OF LEAF LITTER PRODUCTION

% OF VARIABILITY EXPLAINED :

LAXE1	LAXE2
40	26

CORRELATION VARIABLES/AXES :

VARIABLE	LAXE1	LAXE2
C	0.18	0.24
N	-0.21	-0.4
C/N	0.34	0.51
P	-0.29	0.43
K	-0.82	0.04
Na	0.01	-0.33
Ca	-0.95	0
Mg	-0.54	0.17
Mn	0.3	-0.49
Fe	-0.46	-0.1
Al	-0.36	-0.14

Fig. 12.2. Ordination/classification of the leaf nutrient content in the stands (Muys *et al.*, 1992).

Table 12.3. ANOVA of the holorganic layer biomass (g/m^2), where $F = 4.179$ and significance level = 0.0128.

Stand	Average (g/m^2)	Multiple range test (95% confidence interval)	
QPA	727	*	
TIL	136	*	*
PRU	96		*
ALN	63		*
FR1	5		*

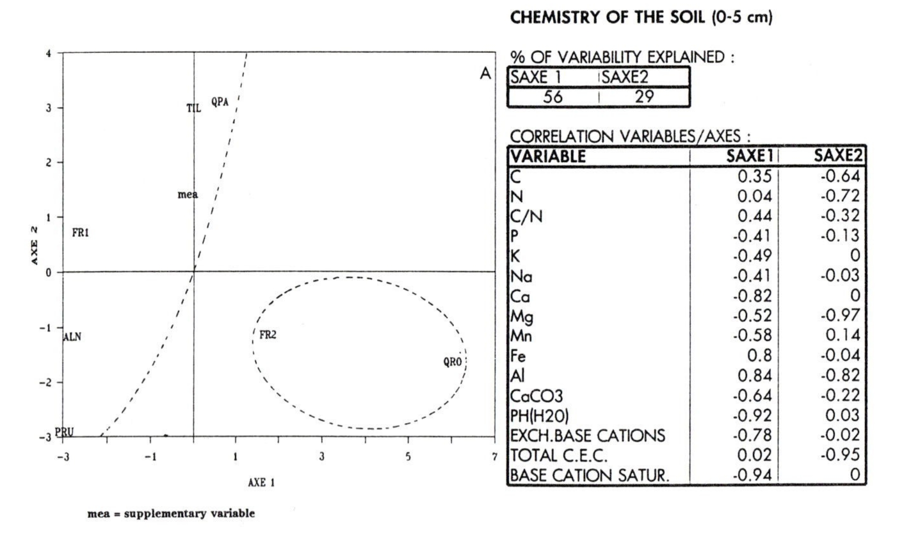

SAXE 1	SAXE2
56	29

VARIABLE	SAXE1	SAXE2
C	0.35	-0.64
N	0.04	-0.72
C/N	0.44	-0.32
P	-0.41	-0.13
K	-0.49	0
Na	-0.41	-0.03
Ca	-0.82	0
Mg	-0.52	-0.97
Mn	-0.58	0.14
Fe	0.8	-0.04
Al	0.84	-0.82
CaCO3	-0.64	-0.22
PH(H20)	-0.92	0.03
EXCH.BASE CATIONS	-0.78	-0.02
TOTAL C.E.C.	0.02	-0.95
BASE CATION SATUR.	-0.94	0

Fig. 12.3. Ordination/classification of the soil nutrient concentration in the stands (Muys *et al.*, 1990).

Prunus, *Fraxinus* and the meadow (989–1334 g/m^2) and distinctly, but not significantly lower values than under *Alnus* and *Tilia* (673–712 g/m^2) (Table 12.5).

The herbaceous vegetation in the stands does not yet reflect the soil changes that have taken place. The present-day meadow still has the same vegetation type as the original meadow, dominated by grasses such as *Dactylis glomerata* and *Holcus lanatus*. The young forest stands all have a vegetation mainly dominated by *Urtica dioica*. The herbaceous above-ground biomass produced yearly is dependent on the amount of light penetrating the canopy and can vary from the almost complete absence of vegetation under the relatively dense canopy of *Tilia*

Table 12.4. ANOVA of the mineral soil pH (H_2O) at 0–5 cm, where $F = 12.998$ and significance level = 0.0000.

Stand	Average	Multiple range test (95% confidence interval)		
QPA	4.97	*		
MEA	5.18	*	*	
TIL	5.26	*	*	
ALN	5.51		*	
PRU	5.54		*	
FR1	6.08			*

Table 12.5. ANOVA of the earthworm biomass (g/m^2), where $F = 6.167$ and significance level = 0.0008.

Stand	Average (g/m^2)	Multiple range test (95% confidence interval)		
QPA	344	*		
ALN	673	*	*	
TIL	712	*	*	
FR1	989		*	*
MEA	1017		*	*
PRU	1334			*

to a maximum 350 g/m^2 under the sparser canopy of *Quercus palustris.* The old stands are dominated by perennial herbs, *Fraxinus* by *Anemone nemorosa* and *Lamium galeobdolon*; *Quercus robur* by *Rubus fruticosus* and *Pteridium aquilinum.*

The species diversity in the herb layer was quite variable. The original meadow contained 14 species, while the untreated meadow today has 20 species, the *Quercus palustris* stand 10, *Alnus* 13, *Prunus* 18, *Fraxinus* 24 and the *Tilia* stand 30. The old *Quercus robur* stand has 7 and the old *Fraxinus* stand has 33. Typical species of woodlands with a mull humus (like under the old *Fraxinus* stand), such as *Dryopteris filix-mas, Arum maculatum, Polygonatum multiflorum, Ficaria verna* and *Adoxa moschatellina,* have started to appear – especially under the young *Prunus* and *Fraxinus* stands. None of the young stands has developed vegetation consisting of acid-loving woodland species as is found in the old *Quercus robur* stand.

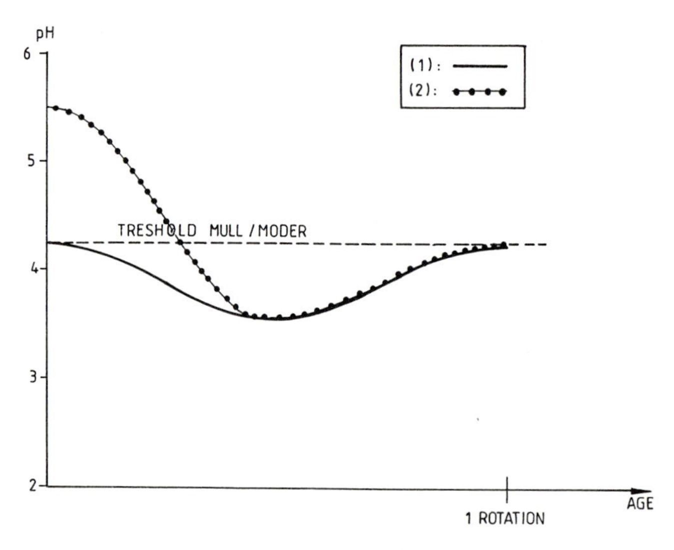

Fig. 12.4. Time sequence graph of the pH evolution following afforestation with tree species having refractory litter. (1) acid initial soil condition (Page, 1962), (2) neutral initial soil condition.

Discussion

After 20 years of forest development on a former meadow, significant differences in several biological and chemical ecosystem variables could be detected under different tree species. In the *Quercus palustris* stand in particular the poor quality of the litter resulted in litter accumulation and was thus the triggering factor leading to the chemical impoverishment of the soil.

A complete recovery of the soil at the end of the rotation, as suggested by Page (1962) and Miller (1988) cannot be expected in soils with a topsoil pH of more than 4.5 but without free $CaCO_3$. The degradation will pass a threshold that is irreversible unless expensive ameliorating measures like liming are carried out (Fig. 12.4). Based on the ordination and classification of 25 forest stands in Flanders (Belgium), Muys and Lust (1992) set threshold at a pH(H_2O) of 4, a C/N-value of 14 and a base cation saturation of 30% in the topsoil (0–5 cm).

The ecological explanation for this threshold is two-fold. First, at this level of acidity all burrowing earthworm species, both endogeic and anecic, disappear. Endogeic earthworms are pale-coloured soil-living earthworms. They ingest humus-rich soil and in this way create horizontal gallery systems in the first 30 cm of mineral soil. Anecic earthworms are dark-pigmented earthworms that build deep, vertical galleries. At night they feed on litter from the soil surface (Bouché, 1972). The gradual decrease in abundance of earthworms slows down litter decomposition and almost stops the mixing of soil from different layers; it establishes a vicious circle of biological, chemical and physical soil degradation. Moreover, once the earthworms have gone, their recolonizing capacity is very restricted. This was illustrated in a 80-year-old maple stand, following a beech rotation, on a rich, loamy soil. Despite the fact that good litter was produced, the soil had not become an active mull humus again because of the lack of burrowing earthworms in the surrounding area (Muys, 1989). Second, the soil passes from the exchange-buffer range into the aluminium-buffer range at a pH of 4.2 (Ulrich, 1983). The more the exchange surfaces are saturated with aluminium plus iron the more their Ca/Al and Mg/Al selectivity decreases and the more difficult it becomes to absorb base cations (Hildebrand, 1991).

These considerations lead us to the conclusion that afforestation of meadows that have an active earthworm community results in the humus developing in one of two ways. The mull humus is maintained if most of the litter produced is of good enough quality; if this is not the case the mull humus develops into a moder humus (Fig. 12.5).

How can these assessments be interpreted if one of the aims of forest management is stability? Van Miegroet (1990) mentions loss of internal regulation over the movement of mineral soil nutrients as one of the main causes for forest degradation. Ulrich (1981, 1983) considers soil acidification and cation leaching as the main threat to forest stability and Wittich (1963) states that mull humus is desirable in all aspects – especially for water economy and nutrient status. Additional factors that must be considered include the continuous external proton loads due to air pollution and the fact that more than three-quarters of the forest area of Flanders is already situated in the aluminum- or iron-buffer range. If all these factors are taken into account it is clear that all tree species that are known to cause soil degradation should be excluded from the afforestation of agricultural land in Flanders. Silviculture on these soils must concentrate on the cultivation of valuable broadleaved species such as ash, maple and cherry. Plantations of fast-growing pioneer species such as poplar, willow and alder are also acceptable.

To what degree trees with leaf litter that takes a relatively long time to decompose, such as *Quercus* spp., can be tolerated in a tree species mixture, given that the mull humus must be maintained, remains an open question and certainly needs more research. It is likely, however, that it depends highly on the initial biomass of anecic earthworms present: anecic-poor mulls such as the *Alnus* stand are much more sensitive to litter accumulation than anecic-rich mulls such as the *Prunus* stand (Muys *et al.*, 1992).

The afforestation of meadowland appears to result in an increase in the species diversity of the ground flora. This is comparable with what Rehfüss *et al.* (1990) found for afforested arable land. The dominance of *Urtica dioica* can be reduced by underplanting a substorey of shade-tolerant shrubs, such as *Corylus avellana, Cornus* spp., *Viburnum opulus, Alnus glutinosa* and so forth. Under these conditions, typical woodland species will begin to colonize the plantation relatively quickly so long as old woodland still exists nearby (see Chapter 3).

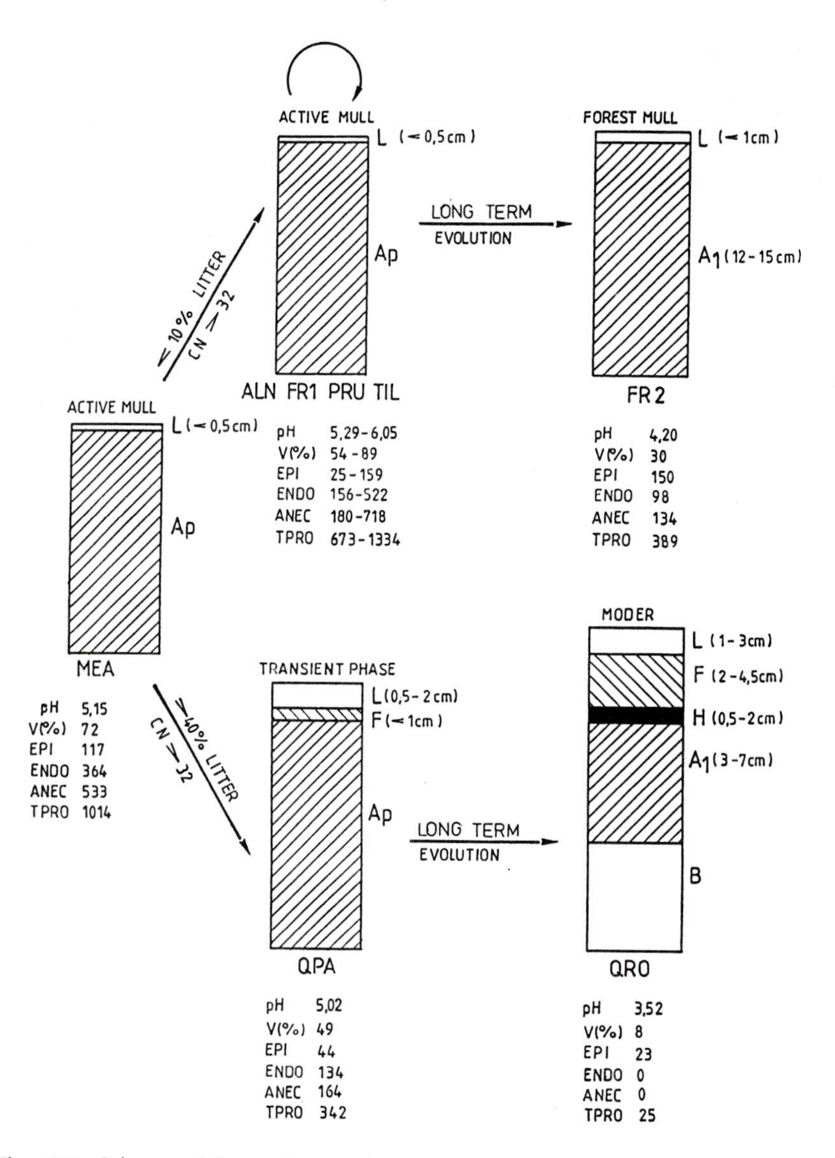

Fig. 12.5. Scheme of the evolution of the earthworm community, the humus type and the soil properties following afforestation of a meadow with different tree species (Muys *et al.*, 1992).

CONCLUSIONS

In summary, our main conclusions are as follows. First, the use of a tree species with litter that takes a long time to decompose in afforestation of meadowland on a sandy loam soil leads to litter accumulation and soil acidification. Second, this degradation is likely to be irreversible when topsoil pH decreases below pH4, as below this threshold earthworms disappear and the soil enters the aluminum-buffer range. Third, forests with mull humus are more resistant to external proton loads than forests with moder humus and hence mull humus is preferable for maintaining forest stability. Fourth, when good-quality soils are afforested, trees with high-quality litter, including valuable broadleaved species such as ash, cherry, maple or walnut, and fast-growing species such as poplars and willows, should be used. Finally, under these species, an interesting vegetation can develop. Typical woodland herbaceous species will appear, depending on the amount of light and the distance from old woodlands.

13

Ecological Restrictions on the Afforestation of Valley Grounds in Flanders, Belgium: Guidelines for Government Policy

W. BUYSSE

INTRODUCTION

On 13 June 1990, a new forest decree came into operation in Flanders. The first executive ordinance under the decree provides grants for afforestation and reforestation by private landowners. Guidelines were set up for the forestry service to decide whether or not a grant should be provided. There are three levels in the decision-making process: first, statutory regulations; second, government policy, and third, the ecological requirements of species and communities. In some valleys, unconsidered afforestation could have a negative impact on the nature-conservation value of sites. The occurrence of specific grassland and marsh plant communities and of some meadowland bird species are used as criteria to judge whether afforestation is appropriate. Special attention has been paid to plantations with clones of poplar. Some guidelines are provided to enhance the ecological value of poplar plantations.

Before discussing these guidelines, it is useful to give an overview of the Belgian constitution and the Flemish Forest Service. Belgium is a country with about 10 million inhabitants and a surface of 30 513 km^2. It is divided into Dutch-speaking, French-speaking and German-speaking parts. State reforms over the last 10 years mean that there is now a national government and a government for each region and community (Table 13.1). At the national level, laws are made by Royal or Ministerial ordinances; on a regional level decrees are made by executive ordinances.

The Flemish Forest Service is now a part of the Department of Nature Conservation and Development. Under the former Belgian Forest Law of 1854, the Belgian Forest Service belonged first to the Administration of Finance and later, from 1885 onwards, it became an independent directorate-general in the department of agriculture. Today, however, the Flemish Forest Service is on the

Table 13.1. The Belgian constitution of the State.

Provinces ($n = 9$)
Linguistic areas ($n = 4$): (Dutch speaking, French speaking, the bilingual area of the capital city Brussels, German speaking)
Communities ($n = 3$) (Flemish community, French community, German-speaking community)
Regions ($n = 3$) (Flemish region, Walloon region, Brussels' capital city region)

Legislature

National (law)
- King
- House of representatives
- Senate

Regional (decree)
- The Flemish council
- The Wallonian regional council (local matters)
- The French community council (cultural and personal matters)
- The council of the German-speaking community
- The council of the capital city Brussels (local matters, ordinances, no decrees)

Executive

National (Royal ordinance or Ministerial ordinance)
- King plus national government

Regional (executive ordinance)
- The Flemish executive
- The executive of the Wallonian region
- The executive of the French community
- The executive of the capital city Brussels

lowest step on the administrative ladder. This emphasizes two points. First, forest issues in Flanders, which has a forest cover of 8.5%, are considered less important than in Belgium as a whole with its forest cover of 20%. Second, the change of department shows that in densely populated and highly industrialized Flanders attention is increasingly directed to social and conservationist issues, while the traditional issues of timber and game production play a less important role.

The new forest legislation that was introduced by the Flemish Forest Decree of June 1990 replaced the Belgian Forest Law of 1854. The decree:

1. Gives a detailed definition of what is meant by the concept of forest.
2. Applies to all forests, both public and private.
3. Recognizes several forest functions:

(a) economic function;
(b) social and educative function;
(c) shelter function;
(d) ecological function;
(e) scientific function.

4. Stresses that forests are multifunctional.
5. Gives detailed instructions regarding the organization of the forest service, forest management and forest protection.

Before this new forest decree was executed, there was a vacuum in the legislation about the private forests. No grants were payable to private woodland owners, and there were no effective regulations to stop excessive fellings. The first Ordinance of the Flemish executive that followed the Forest Decree provided grants for private woodland owners who opened their property to the public; who made a common management plan together with owners of adjacent properties, and who made plantations or re-established woodlands. The species that are subsidized are listed in Table 13.2.

THE AMELIORATION OF THE ECOLOGICAL VALUE OF PLANTATIONS

The proportion of Flanders that is wooded is very low (8.5%) and much of this woodland is underproductive in terms of economic, social and ecological benefits. The Forest Decree, therefore, which stresses the multifunctional aspect of forests and provides grants for afforestation, may well be a first step towards the enlargement of the forest area and the improvement of the different uses to which the woodland is put.

However, afforestation can have a negative effect on the ecological value of particular sites – particularly grassland. Many of these grassland communities have been destroyed over the last 10 years by intensive agricultural land management. A private forest owner with little knowledge about forest management and who is mostly only interested in the financial benefits, will use simple planting techniques and may not make a suitable choice of tree species. In valley areas, the most popular choice will be some sort of poplar clones; this is acceptable from both an economic and social point of view.

The ecological effects of establishing poplars depend to a great extent on the establishment techniques used. If the site is ploughed, manured and drained before the poplars are planted then this will clearly result in a heavily modified vegetation. However, if suitable techniques are used, plantations of poplars can be made more ecologically acceptable.

The ground vegetation of most poplar plantations consists of common species that indicate rich, disturbed soils such as stinging nettle (*Urtica dioica*) and brambles (*Rubus* spp.). The 15 species that are most strongly associated with

Table 13.2. Tree species that can be subsidized.

I. 100 000 Bfr/ha
- *Quercus robur*
- *Quercus petraea*
- *Fraxinus excelsior*

II. 80 000 Bfr/ha
- *Fagus sylvatica*
- *Prunus avium*
- *Carpinus betulus*
- *Acer campestre*
- *Acer pseudoplatanus*
- *Acer platanoides*
- *Tilia cordata*
- *Tilia plathyphyllos*
- *Tilia × vulgaris*
- *Ulmus glabra*
- *Ulmus minor*

III. 60 000 Bfr/ha
- *Quercus palustris*
- *Quercus rubra*
- *Castanea sativa*
- *Juglans regia*
- *Alnus glutinosa*
- *Betula pendula*
- *Betula pubescens*
- *Salix alba*
- *Salix fragilis*
- *Salix × rubens*
- *Populus nigra*
- *Populus alba*
- *Populus tremula*
- *Populus canescens*
- *Pinus sylvestris*

IV. 40 000 Bfr/ha
- *Robinia pseudoacacia*
- *Taxus baccata*
- *Juniperus communis*
- *Pinus nigra* var. *corsicana*
- *Pseudotsuga menziesii*
- *Larix kaempferi*
- *Larix × eurolepis*
- *Alnus incana*
- Poplar clones mixed with indigenous tree species or shrubs

V. 20 000 Bfr/ha
- Poplar clones

Populus × *euramericana* are given in Table 13.3 (Hermy, 1985). There is, however, no evidence that this is due to the fact that poplar clones produce a large number of leaves that contain a high nitrogen level. The leaves decompose quickly, so that the next year one can hardly see their remains in the litter; however, the amount of foliage is comparable with other tree species (Bray and Gorham, 1964), as is the nitrogen level (Fig. 13.1).

The species in Table 13.3 are strong competitors that can quickly colonize a site following disturbance. This happens also when other tree species are planted on former agricultural land (Noirfalise, 1969). Most tree species are planted more densely than poplars, however, and so the amount of light available is greatly reduced in comparison. One of the most important factors influencing the occurrence of these common species will be the amount of light. This can also be concluded from the fact that there is no increase in species richness of the herb and moss layers when an existing coppice wood is planted up with poplar clones at a spacing of 8 × 8 m or 12 × 12 m (Verlinden, 1987). Other important factors will be disturbance in the broad sense (e.g. from ploughing, wheel tracks, drainage), the richness of the soil (or in other words the former land-use) and the quality of the air (N-deposition).

To enhance the ecological value of a poplar plantation the trees should be spaced widely (at least 9 × 9 m) and an underwood of broadleaved trees or shrubs should be planted at a spacing of 2 × 2 m. The soil and the water economy of the site should be disturbed as little as possible. Hence no ploughing, manuring nor drainage of the site should take place.

Table 13.3. List of the most strongly associated species with *Populus* × *euramericana*.

Species
Adoxa moschatellina
Angelica sylvestris
Cirsium palustre
Filipendula ulmaria
Galeopsis tetrahit
Galium aparine
Geum urbanum
Glechoma hederacea
Moehringia trinervia
Poa trivialis
Ranunculus ficaria
Stachys sylvatica
Urtica dioica
Brachytecium rutabulum
Eurynchium praelongum

Souce: Hermy, 1985.

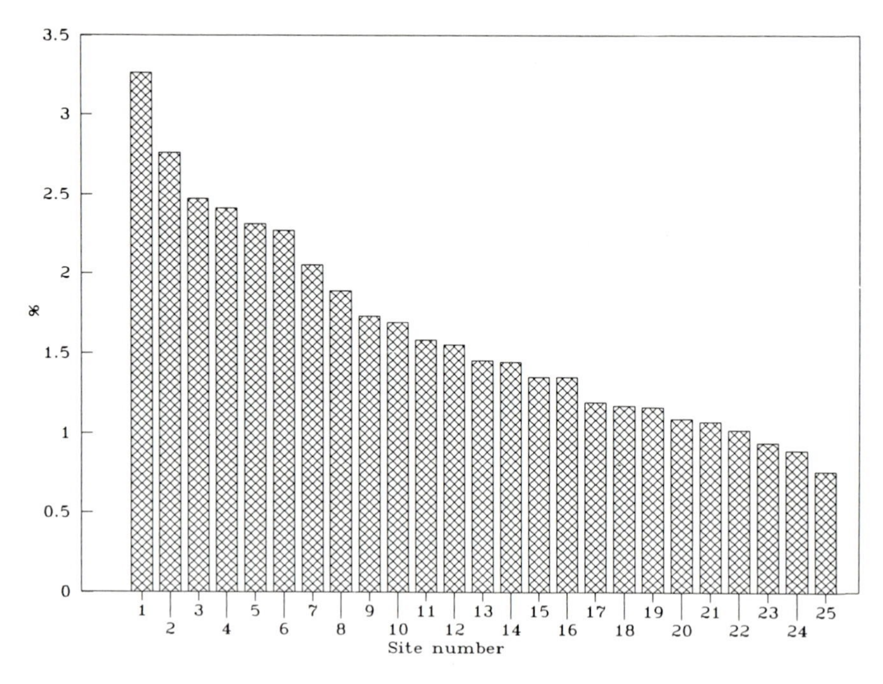

Fig. 13.1. N-level of the leaf litter (%). (See Table 13.4 for details.) (Source: Muys *et al.*, 1989b.)

If these guidelines are followed there are various advantages. The economic advantages are:

1. A lesser number of trees at the start provides a higher yield (Faber, 1990).
2. A plant distance of 9 × 9 m results in the highest basal area (Lust and Van Gijsel, 1986).
3. A wide planting distance makes thinning unnecessary and diminishes the risk of diseases (De Kam, 1990; Heybroek and Schmidt, 1990).
4. The underwood can be exploited as coppice.
5. The underwood protects the poplar plantation against unfavourable climatological circumstances.

The ecological advantages are:

1. A reduction in the development of common species in the herb layer.
2. Woodland species can expand more quickly.
3. The ornithological value increases. A poplar plantation with trees older than 40 years and with an underwood provides similar habitat for birds to a natural oakwood (Van Hees, 1978).
4. The site will already have developed some 'natural' woodland charactersitics when the poplars are felled.

Table 13.4. List with the numbers of the sites shown in Figure 13.1.

1.	*Alnus incana*
2.	*Fraxinus excelsior*
3.	*Fraxinus excelsior*
4.	*Fagus sylvatica*
5.	*Pinus sylvestris*
6.	*Acer pseudoplatanus*
7.	*Populus × euramericana*
8.	*Tilia platyphyllos*
9.	*Fagus sylvatica*
10.	*Populus × euramericana*
11.	*Quercus robur*
12.	*Pinus nigra calabrica*
13.	*Quercus robur*
14.	*Fagus sylvatica*
15.	*Quercus robur*
16.	*Quercus robur*
17.	*Quercus robur*
18.	*Pinus sylvestris*
19.	*Prunus avium*
20.	*Fagus sylvatica*
21.	*Fagus sylvatica*
22.	*Quercus palustris*
23.	*Fagus sylvatica*
24.	*Fagus sylvatica*
25.	*Pinus nigra calabrica*

In addition, many people consider that such a plantation with underwood produces a woodland landscape that is richer and more 'natural' in appearance (Stoffel, 1985).

GUIDELINES FOR THE PROVISION OF GRANTS

There are five main grant specifications.

- no grant;
- grant for indigenous tree species;
- grant for poplar clones or exotic tree species with an underwood of indigenous tree species and with the conservation of the existing woody vegetation. The planting distances are specified. The next generation of the stand must be made by selection of the indigenous tree species from the underwood;
- grants for poplar clones or exotic tree species with the establishment or conservation of a shrub layer; and
- no specifications.

In the first four classes, structural works such as drainage and ploughing of the site are restricted or not allowed. When deciding which class of grant is applicable to a particular afforestation scheme, account has to be taken of the statutory regulations, government policy and the ecological requirements of species and communities.

Statutory regulations

These fall into various groups. One group regulates which local departments, services and officials should be consulted for advice and permissions associated with afforestation. A second group consists of laws that have an impact on specific areas such as the Ramsar convention and the EC bird directive. The most important of this group is a Royal Ordinance of 1972 that divides the country into land-use zones including areas for nature conservation and forestry. The amount of subsidy available for a particular site will depend on the location of the site in relation to these zones and specially designated areas.

Government policy

A policy declaration of the minister of the Flemish Community of the Environment, Nature Protection and Land Planning (1990) developed the idea of various policy categories. These included the 'Green Headstructure', the Ecological Infrastructure, Nature in the Urban Sphere and Geographical Values. The Green Headstructure is divided into Nature Protection Areas (*natuurkerngebieden*, literally translated: Nature Nucleus Areas), Nature Development Areas, Nature Connection Areas and Nature Buffer Areas. The division is based mainly upon the occurrence of nature reserves and nature values. At present there is no legislation to back these concepts legally; it is just policy.

Although the development of policies like this can be an important means of stopping the further degradation of the countryside, there are some fundamental shortcomings. The scientific foundation is very weak. Much is based on the island biogeography theory of MacArthur and Wilson (1967) which has never been proven unambiguously for continental situations. Another important base for the Green Headstructure is the Biological Evaluation Map of Belgium (De Blust *et al.*, 1985). This gives an overview of the Belgian vegetation and indicates 'biological valuable' (light green) and 'biological very valuable' (dark green) areas. As an inventory map it has its merits, but some of the evaluation on which it is based can be criticized for being subjective. One negative result of the Green Headstructure is that it has come to be used more as a device in a battle between the different groups that make use of the countryside than as a means of protecting the countryside.

Ecological requirements of species and communities

THE FLORA OF GRASSLANDS AND MARSHES

No afforestation grants will be given for marshland or for the following types of grassland: *Calthion, Molinion caeruleae* and *Arrhenaterion elatioris* (symbols Hc, Hm and Hu, respectively, on the Biological Evaluation Map). Clearly, anyone proposing to plant trees should consult the map to check the area concerned does not fall into one of these categories. In addition, a site visit is recommended because the grassland could have changed since the survey was carried out or the survey may have been superficial.

THE FLORA OF FORESTS

In ancient woods, a grant for reforestation will only be provided if indigenous tree species are used and if no drainage works are undertaken. This regulation should help to protect woodland plants that are usually unable to cope with grazing, soil disturbance or competition with tall herbs (Rackham, 1980).

AVIFAUNA

The afforestation of open valley grounds will mainly affect stilt-walkers and other meadowland bird species. Planting will reduce the availability of breeding sites and bring about a higher predation of birds and eggs by small mammals and crows. In areas mentioned in the EC bird directive, and in kilometre squares of regional breeding bird inventories where *Gallinago gallinago, Limosa limosa, Anas querquedula* and *Saxicola rubetra* are found, a grant will only be provided if the planting is sited next to existing woodland.

CONCLUSIONS

It is difficult to implement ecological issues into legislation and policy. Laws have to be clearly defined and unambiguous, while in ecology rules are complicated and there are frequent exceptions. The situation is made more difficult by the relatively fixed points of view of various interest groups; in particular the wood production and conservation lobbies. An additional problem is that there is often a lack of recent data on distibution of different types of habitat and species that might be adversely affected by afforestation.

The rigid application of afforestation policy based on a rigid zonation of the countryside is not flexible enough. A more pragmatic and realistic approach, given the long-term impacts of afforestation, the relatively short-term influence of policy and the relative freedom of private landowners and farmers to change land use, is to make afforestation decisions on a case by case basis.

References

Allott, N.A., Mills, W.R.P., Dick. J.R.W., Eacrett, A.M., Brennan, M.T., Clandillon, S., Phillips, W.E.A., Critchley, M. and Mullins, T.E. (1990) *Acidification of Surface Waters in Connemara and South Mayo – Current Status and Causes.* duQuesne Ltd, Dublin.

Anderson, M.A. (1979) The development of plant habitats under exotic forest crops. In: Wright, S.E. and Buckley, G.P. (eds) *Ecology and Design in Amenity Land Management.* Recreation Ecology Research Group, Wye College, Wye, pp.87–109.

Angeloni, G.A. (1885) *Atti della Giunta per l'Inchiesta agraria sulle condizioni della classe agricola.* Vol.XII. Forzani e C. Tipografia del Senato, Roma, pp.115–116.

Anko, B. (1988) The changing role of forest in the karst landscape of Slovenia, Yugoslavia. In: Salbitano, F. (ed.) *Human Influence on Forest Ecosystems Development in Europe.* Pitagora Editrice Bologna, Bologna, pp.95–108.

Ascacibar, M. and Iturrondobeitia, J.C. (1984) Estudio de las poblaciones de oribatidos en tres medios urbanos de la ciudad de Bilbao. Aplicacion del modelo log-lineal de Motomura. *Kobie* **13**, 159–160.

Ascari, M.C. (1938) Topografia di Genova-Quarto. *Genova-Rivista Municipale* **12**(**1**), pp.19–26.

Atlas of Ireland (1978). Royal Irish Academy, Dawson Street, Dublin 2.

Avery, M.I., Winder, F.W.R. and Egan, V. (1989) Predation on artificial nests adjacent to forestry plantations in northern Scotland. *Oikos* **55**, 321–323.

Avery, M.I. (1989) Effects of upland afforestation on some birds of the adjacent moorland. *Journal of Applied Ecology* **26**, 957–966.

Avery, M.I. and Haines-Young, R.H. (1990) Population estimates for the dunlin *Calidris alpina* derived from remotely sensed satellite imagery of the Flow Country of northern Scotland. *Nature* **344**, 860–862.

Avery, M.I. and Leslie, R. (1990) *Birds and Forestry.* Poyser, London.

Babel, U. and Benckiser, G. (1975) Ökologische Untersuchungen an aufgeforsteten Weideflächen. *Daten und Dokumente zum Umweltschutz* **14**, 31–37.

Bainbridge, I.P., Minns, D.W., Housden, S.D. and Lance, A.N. (1987) *Forestry in the*

Flows of Caithness and Sutherland. RSPB, Sandy.
Batterbee, R.W. (1990). The causes of lake acidification, with special reference to the role of acid deposition. *Philosophical Transactions of the Royal Society, London,* B **327**, 339–347.
Bayes, K. and Henderson, A. (1988) Nightingales and coppiced woodland. *RSPB Conservation Review* **2**, 47–49.
Bellini, D. (1879) *Monografia agricola industriale del Circondario di Campobasso.* Colliti, Campobasso, p.22.
Banerjee, S. (1984) Qualitative and quantitative composition of Oribatid mites *Acarina* in relation to certain soil factors. In: Griffiths, D.E. and Bowman, C.D. (eds) *Acarologia VI.* Ellis Horwood, Chichester, pp.878–885.
Bibby, C.J. (1986) Merlins in Wales: site occupancy and breeding in relation to vegetation. *Journal of Applied Ecology* **23**, 1–22.
Bibby, C.J. (1987) Foods of breeding merlins (*Falco columbarius*) in Wales. *Bird Study* **34**, 64–70.
Biondi, E. and Ballelli, S. (1973) Osservazioni su due rimboschimenti a *Pinus nigra* ssp. *nigra* nella zona di Fabriano (Marche). *Archivo Botanico e Biogeografico Italiano* IL, iv s, 18 (3–4), 163–171.
Bird, S.C., Walsh, R.P.D. and Littlewood, I.G. (1990) Catchment characteristics and basin hydrology: their effects on streamwater acidity. In: Edwards, R.W., Gee, A.S. and Stoner, J.H. (eds) *Acid Waters in Wales.* Kluwer Academic Publishers, Dordrecht, pp.203–221.
Birks, H.J.B. (1988) Long-term change in the British uplands. In: Usher, M.B. and Thompson, D.B.A. (eds) *Ecological Change in the Uplands.* Blackwell Scientific Publications, Oxford, pp.37–56.
Bishop, K.D. (1990) "Multi-purpose woodlands in the countryside around towns: fact or fiction?" PhD Thesis, University of Reading.
Bouché, M.B. (1972) *Lombriciens de France; Ecologie et Systématique.* INRA, Paris.
Bowden, C. and Hoblyn, R. (1990) The increasing importance of restocked conifer plantations for woodlarks in Britain: implications and consequences. *RSPB Conservation Review* **4**, 25–31.
Boycott, A.E. (1934) The habitats of land Mollusca in Britain. *Journal of Ecology* **22**, 1–38.
Braun-Blanquet, J. (1964) *Pflanzensoziologie,* 3 Aufl., Wien.
Bray, J.R. and Gorham, E. (1964) Litter production in the forests of the world. *Advances in Ecological Research* **2**, 101–157.
Bright, P.W. and Morris, P.A. (1990) Habitat requirements of dormice *Muscardinus avellanarius* in relation to woodland management in south west England. *Biological Conservation* **54**, 307–326.
Brotherton, I. (1986) Agricultural and afforestation controls: conservation and ideology. *Land Use Policy* **January**, 21–30.
Brouwer, F.M. and Chadwick, M.J. (1991) Future land-use patterns in Europe. In: Brouwer, F.M., Thomas A.J., and Chadwick M.J., (eds) *Land Use Changes in Europe.* Kluwer Academic Publishers, Dordrecht, pp.1–20.
Brower, J. and Zar, J.H. (1977) *Field and Laboratory Methods for General Ecology.* William. C. Brown Publishers, Dubuque, Iowa.
Brown, A.H.F. and Oosterhuis, L. (1981) The role of buried seed in coppicewoods. *Biological Conservation* **21**, 19–38.
Brown, A.H.F. and Warr, S.J. (1992) The effects of changing management on seed banks

in ancient coppices. In: Buckley, G.P. (ed.) *Ecology and Management of Coppice Woodlands.* Chapman and Hall, London, pp.147–166.

Brown, J.L. (1969) Territorial behaviour and population regulation in birds. A review and re-evaluation. *The Wilson Bulletin* **81**, 293–329.

Brückner, J., Rehfüss, K. and Makeschin, F. (1987) Braunerden auf Schotterrassen im Alpenvorland unter Grünland, Fichten-Erstaufforstung, Laubbaum-Folgebestand und altem Wald. Beitrag zum Studium der Einflüsse verschiedenartiger Landnutzung auf Bodeneigenschaften. *Mitteilungen des Vereins für Forstliche Standortskunde und Forstpflanzenzüchtung* **33**, 49–61.

Buckley, G.P. (ed.) (1989) *Biological Habitat Reconstruction.* Belhaven Press, London.

Buckley, G.P. (ed.) (1992) *Ecology and Management of Coppice Woodlands.* Chapman and Hall, London.

Bull, K.R. and Hall, J.R. (1987) Aluminium fractionation in freshwaters. In: Rowland, A.P. (ed.) *Chemical Analysis in Environmental Research.* Institute of Terrestrial Ecology, Edinburgh, pp.6–12.

Bundy, G. (1979) Breeding and feeding observations on the black-throated diver. *Bird Study* **26**, 33–36.

Buysse, W. (1991) "Ecologische randvoorwaarden voor bebossen en herbebossen van valleigronden". (Ecological restrictions on the afforestation and reforestation of valley grounds.) Unpublished report, University of Ghent, Laboratory of Forestry.

Campbell, L.H. and Talbot, T.R. (1987) The breeding status of black-throated diver (*Gavia arctica*) in Scotland. *British Birds* **80**, 1–8.

Cancela Da Fonseca, J.P. (1969) L'outil statistique en biologie du sol. VI. Théorie de l'information et diversité spécifique. *Revue de Ecologie et de Biologie du Sol* **6**, 533–555.

Coineau, Y. (1974) *Introduction a l'étude des microarthropodes du sol et son annexes.* Doin.

Corbellini, R. (1986) Per un repertorio delle fonti catastali dell'ottocento. *Metodi e Ricerche* **2**, 51–85.

Cottrill, S.M. (1988) Spatial distribution of wet and dry sulphur deposition in the United Kingdom. In: Mathy, P. (ed.) *Air Pollution and Ecosystems.* D. Reidel Publishing, Dordrecht, pp.493–498.

Countryside Commission (1989a) *A New National Forest in the Midlands: A Consultation Document.* Countryside Commission, Cheltenham.

Countryside Commission (1989b) *Forests for the Community.* Countryside Commission, Cheltenham.

Countryside Commission (1991) *Nature Conservation in a Community Forest: Guidelines for Thames Chase.* Countryside Commission, Cheltenham.

Cremaschi, M., Ferraris, M., Maggi, R. and Ottomano, C. (1992) Case Cordona: da bosco a campo durante l'eta del Bronzo. In: Maggi R. (ed.) Archeologia preventiva lungo il percorso di un metanodotto. *Quaderni della Soprintendenza Archeologica della Liguria* 4, Chapter 3.

Crisp, D.T. (1989) Some impacts of human activities on trout, *Salmo trutta*, populations. *Freshwater Biology* **21**, 21–33.

Croce, G.F. and Moreno, D. (1988) The geographical effects of forestry law 20/VII/1877 in Liguria (N.W. Italy). In: Salbitano, F. (ed.) *Human Influence on Forest Ecosystem Development in Europe.* Pitagora Editrice, Bologna, pp.311–319.

CTR (1979) Regione Liguria *Carta Tecnica Regionale.* Sezioni 214130; 231010.

Cuadras C.M. (1981) *Analisis Multivariante.* Eunibar, Barcelona.

Cunningham, I. (1991) "Forestry Expansion: A Study of Technical, Economic and Ecological Factors". Unpublished report, Forestry Commission, Edinburgh.

Daget, J. (1979) *Les modelles mathématiques en écologie.* Masson, Paris.

Davies, M. (1988) The importance of Britain's twites. *RSPB Conservation Review* **2**, 91–94.

Davison, W. (1990) A practical guide to pH measurement in freshwaters. *Trends in Analytical Chemistry* **9** (**3**), 80–83.

De Blust, *et al.* (ed.) (1985) *Biologische waarderingskaart van België. Algemene verklarende tekst.* (Biological Evaluation Map of Belgium. Explanatory text.) Die Keure, Brugge.

De Kam, M. (1990) Het effect van de plantafstand op de ontwikkeling van ziekten. (The effects of planting width on the development of diseases.) *Nederlands Bosbouwtijdschrift* **62**(**7**), 236–239.

De Coninck, E. (1972) Bijdrage tot de vergelijkende oecologische studie van de Coleopterenfauna van een bos en een weide, te Gontrode. *Licentiaatsverhandeling RUGent*, 7–12.

Del Re, G. (1836) *Descrizione Topografica Fisica Economica Politica de'Reali Dominj al di qua del Faro nel Regno delle Due Sicilie.* Tomo III, Tipografia Dentro La Pietà De Turchini, Napoli, pp.16–66.

Di Martino, P. (1986) 'Pascoli boscosi del Molise' Pratiche silvo-pastorali nella foresta di Montedimezzo (XVII–XIX Secolo). *Quaderni Storici nuova serie 62.* Il Mulino, Bologna.

Di Martino, P. (1988) Silver fir in three forests of Molise. In Salbitano, F. (ed.) *Human Influence on Forest Ecosystems Development in Europe.* Pitagora Editrice, Bologna, pp.335–338.

Dougan, W.K. and Wilson, A.L. (1974) The absorptiometric determination of aluminium in water. A comparison of some chromogenic reagents and the development of an improved method. *Analyst* **99**, 413–430.

Driscoll, C.T. (1980) "Chemical characterisation of some acidified lakes and streams in the Adirondack Region of New York State". PhD Thesis, Cornell University, Ithaca, New York (not available but described in detail in, for example, Bull and Hall, 1987).

Dzwonko, Z. and Loster, S. (1989) Distribution of vascular plant species in small woodlands on the western Carpathian foothills. *Oikos* **56**, 77–86.

Edwards, R.W., Gee, A.S. and Stoner, J.H. (eds) (1990) *Acid Waters in Wales.* Kluwer Academic Publishers, Dordrecht.

Ellenberg, H. (1954) *Landwirtschaftliche Pflanzensoziologie.* I. Unkrautgemeinschaften als Zeiger fur Klima und Boden. Ulmer.

Ellenberg, H. (1988) *Vegetation Ecology of Central Europe.* Cambridge University Press, Cambridge.

Eriksson, M.O. (1984) Acidification of lakes: effects on waterbirds in Sweden. *Ambio* **13**, 260–262.

Essex, S.J. (1990) Afforestation around reservoirs: a case study of the Derwent Valley Water Board, 1936–1973, *Applied Geography* **10**(**2**), 111–124.

Evans, P.R. and Pienkowski, M.W. (1984) Population dynamics of shorebirds. In: Burger, J. and Olla, B.L. (eds) *Behaviour of Marine Animals*, Vol.5, pp.83–123.

Faber, P.J. (1990) De invloed van plantafstanden en dunningstijdstippen op de groei en houtproduktie van populier. (The influence of planting width and period of thinning on the growth and production of wood from poplar.) *Nederlands Bosbouwtijdschrift* **62**(7), 226–230.

Ferris-Kaan, R. (1991) *Edge Management in Woodlands.* Forestry Commission (Occasional Paper 28), Edinburgh.

Feruglio, D. (1905) I prati di monte nelle 'Prealpi giulie occidentali' *Bull. Associaizone Agraria Friulana,* Udine, serie V, vol. XXII, pp.286–315.

Flanagan, P.J. (1988) *Parameters of Water Quality. Interpretation and Standards.* Environmental Research Unit, Dublin.

Flegg, J.J.M. and Glue, D.E. (1975) The nesting of the ring ouzel. *Bird Study* **22**, 1–8.

Ford, E.D., Malcolm, D.C. and Atterson, J. (1979) *The Ecology of Even-aged Plantations,* Institute of Terrestrial Ecology, Cambridge.

Forestry Commission (1985) *The Policy for Broadleaved Woodland.* Forestry Commission, Edinburgh.

Forestry Commission (1988) *Forests and Water Guidelines.* Forest Commission, Edinburgh.

Forestry Commission (1989) *Native Pinewoods Grants and Guidelines.* Forestry Commission, Edinburgh.

Foster, C.W. and Longley, T. (1924) *The Lincolnshire Domesday and the Lindsey Survey.* Lincoln Record Society, 19. Morton and Son, Horncastle.

Francis, J.L., Morton, A.J. and Boorman, L.A. (1992) The establishment of ground flora species in new woodland. *Aspects of Applied Biology* **29**, 171–178.

Fretwell, S.D. and Lucas, H.L. (1969) On territorial behaviour and other factors influencing habitat distribution in birds. I. Theoretical development. *Acta Biotheoretica* **19**, 16–36.

Fuller, R.J. and Moreton, B.D. (1987) Breeding bird populations of Kentish sweet chestnut (*Castanea sativa*) coppice in relation to age and structure of coppice. *Journal of Applied Ecology* **24**, 13–27.

Fuller, R.J. and Warren, M.S. (1990) *Coppiced Woodland.* Nature Conservancy Council, Peterborough.

Fuller, R.J. and Warren, M.S. (1991) Conservation management in ancient and modern woodlands: response of fauna to edges and rotations. In: Spellerberg, I.F., Goldsmith, F.B. and Morris, M.G. (eds) *The Scientific Management of Temperate Communities for Conservation.* Basil Blackwell, Oxford, pp.445–471.

Gasson, R. and Hill, P. (1990) *An Economic Evaluation of the Farm Woodland Scheme.* Wye College, University of London.

Gentile, S. (1984) Zonazione altitudinale della vegetazione in Liguria. *Lavori Societa Italiana Biogeografia* **9**, 1–19.

Gentilli J. (1964) *Il Friuli. I climi.* Camera di Commercio Industria e Artigianato, Udine.

Giacobbe, A. (1942) *Il Pino Marittimo.* S. A. Ed. D. Alighieri, Genova-Roma.

Gill, C. (ed.) (1970) *Dartmoor: A New Study.* David and Charles, Newton Abbot.

Goldsmith, F.B. and Wood, J.B. (1983) Ecological effects of upland afforestation. In: Warren, A. and Goldsmith, F.B. (eds) *Conservation in Perspective.* Wiley, London, pp.287–311.

Gobierno Vasco (1986) *Inventario forestal de la Comunidad Autonoma Vasca.* Departamento de Agricultura y Pesca, Gobierno Vasco.

Gomersall, C.H., Morton, J.S. and Wynde, R.M. (1984) Status of breeding red-throated divers in Shetlands, 1983. *Bird Study* **31**, 223–229.

Good, J.E.G. (1987) *Environmental Aspects of Plantation Forestry Wales.* Institute of Terrestrial Ecology (Symposium No. 22), Grange-over-Sands.

Good, J.E.G., Williams, T.G., Wallace, H.L., Buse, A. and Norris, D.A. (1990) "Nature conservation in upland conifer forests". Unpublished report to the Forestry Commission and Nature Conservancy Council, Peterborough.

Goss-Custard, J. (1981) Role of winter food supplies in the population ecology of common British wading birds. *Verhandlungen Ornithologisthen Gesellschaft in Bayern* **23**, 125–146.

Granata, E. (1839) *Elementi di agronomia e della scienza silvana.* Nobile, Napoli.

Grime, J.P. (1987) Dominant and subordinate components of plant communities: implications for succession, stability and diversity. In: Gray, A.J., Crawley, M.J. and Edwards, P.J. (eds) *Colonization, Succession and Stability.* Blackwell Scientific Publications, Oxford, pp.413–428.

Grime, J.P., Hodgson, J.G. and Hunt, R. (1988) *Comparative Plant Ecology.* Unwin Hyman, London.

Groome, H. (1991) Conflicts caused by imbalances in forest policy and practice in the Basque Country. In: Gilg, A.W. *et al.* (eds) *Progress in Rural Policy and Planning,* Vol.1. Belhaven Press, London, pp.140–151.

Guidi, G. (1985) I rimboschimenti di Pino nero in Molise. *Annali Istituto Sperimentale per la Selvicoltura di Arezzo.* Vol.XIV. ISPSA, Arezzo, pp.242–243.

Haeggstrom, C.-A. (1988) Protection of wooded meadows in Aland – problems, methods and perspectives. *Oulanka Reports* **8**, 88–95.

Haeggstrom, C.-A. (1990) The influence of sheep and cattle grazing on wooded meadows in Aland, SW Finland. *Acta Botanica Fennica* **141**, 1–28.

Hagvar, S. (1984) Six common mites species *Acari* in Norwegian coniferous forest soils: Relation to vegetation types and soil characteristics. *Pedobiologia* **27**, 355–364.

Haines, T.A. (1981) Acid precipitation and its consequences for aquatic ecosystems – A review. *Transactions of the American Fish Society* **110**, 609–707.

Haines-Young, R. and Ward, N. (1991) GIS in the development of tree and forestry strategies. *Mapping Awareness '91.* Miles Arnold, Oxford, pp.165–175.

Hallbacken, L. and Tamm, C.O. (1986) Changes in soil acidity from 1927 to 1982–1984 in a forest area of south-west Sweden. *Scandinavian Journal of Forest Research* **1**, 219–232.

Harding, P.T. and Rose, F. (1986) *Pasture Woodlands in Lowland Britain.* Institute of Terrestrial Ecology, Huntingdon.

Harper, J.L. (1977) *Population Biology of Plants.* Academic Press, London.

Harriman, R. and Morrison, B.R.S. (1982) Ecology of streams draining forested and non-forested catchments in an area of central Scotland subject to acid precipitation. *Hydrobiologia* **88**, 251–263.

Harriman, R., Morrison, B.R.S., Caines, L.A., Collen, P. and Watt, A.W. (1987) Long-term changes in fish populations of acid streams and lochs in Galloway south west Scotland. *Water, Air and Soil Pollution* **32**, 89–112.

Harriman, R. and Wells, D.E. (1985) Causes and effects of surface water acidification in Scotland. *Water Pollution Control* **84**, 215–224.

Harris, L.D. (1984) *The Fragmented Forest.* University of Chicago Press, Chicago.

Harris, M.P. (1970) Territory limiting the size of the breeding population of the oyster-catcher (*Haematoqus ostralegus*) – A removal experiment. *Journal of Animal Ecology* **39**, 707–713.

Harrison, A., Tranter, R. and Gibbs, R. (1977) *Landownership by Public and Semi-public Institutions in the UK*, CAS Paper 3. Centre for Agricultural Strategy, University of Reading.

Helliwell, D.R. (1976) The effects of size and isolation on the conservation value of wooded sites in Britain. *Journal of Biogeography* 3, 47–53.

Hermy, M. (1985) "Ecologie en fytosociologie van oude en jonge bossen in Binnen-Vlaanderen". (Ecology and phytosociology of old and young forests in Inner Flanders.) PhD Thesis, University of Ghent.

Hermy, M. and Stieperaere (1981) An indirect gradient analysis of the ecological relationships between ancient and recent riverine woodlands to the south of Bruges (Flanders, Belgium). *Vegetatio* **44**, 43–49.

Heybroek, H.M. and Schmidt, P. (1990) Sommige populiereklonen verdragen geen dichte stand. (Some poplar clones do not tolerate a close planting distance.) *Nederlands Bosbouwtijdschrift* **62**(7), 239–243.

Hildebrand, E. (1991) The influence of forest site fertilization on soil solution chemistry. *FAO/ECE/ILO Seminar on forest site conservation and improvement for sustained yield.* München, pp.193–204.

Hill, M.O. (1979) The development of a flora in even-aged plantations. In: Ford, E.D., Malcolm, D.C. and Atterson, J. (eds) *The Ecology of Even-aged Plantations.* Institute of Terrestrial Ecology, Cambridge, pp.175–192.

Hill, M.O. (1986) Ground flora and succession in commercial forests. In: Jenkins, D. (ed.) *Trees and Wildlife in the Scottish Uplands.* Institute of Terrestrial Ecology, Cambridge, pp.71–78.

Holmes, R.T. (1970) Differences in population density, territoriality and food supply of dunlin in arctic and subarctic tundra. In: Watson, A. (ed.) *Animal Populations in Relation to their Food Resources*, pp.303–317.

Horn, H.S. (1966) Measurement of 'overlap' in comparative ecological studies. *American Naturalist* **100**, 419–424.

Hornung, M., Reynolds, B., Stevens, P.A. and Hughes, S. (1990a). Water-quality changes from input to stream. In: Edwards, R.W., Gee, A.S. and Stoner, J.H. (eds) *Acid Waters in Wales.* Kluwer Academic Publishers, Dordrecht, pp.223–240.

Hornung, M., Le-Grice, S., Brown, N. and Norris, D. (1990b) The role of geology and soils in controlling surface water acidity in Wales. In: Edwards, R.W., Gee, A.S. and Stoner, J.H. (eds) *Acid Waters in Wales.* Kluwer Academic Publishers, Dordrecht, pp.55–66.

Hoskins, W.G. (1955) *The Making of the English Landscape.* Hodder and Stoughton, London.

House of Commons (1990) Agriculture Committee, 2nd Report *Landuse and Forestry*, Vol.1, (Session 1989–90). HMSO, London.

Hudson, P.J. (1988) Spatial variations, patterns and management options in upland bird communities. In: Usher, M.B. and Thompson, D.B.A. (eds) *Ecological Change in the Uplands.* Blackwell Scientific Publications, Oxford, pp.381–397.

Huhta, V. (1979) Evaluation of different similarity indices as measures of succession in arthropod communities of the forest floor after clear cutting. *Oecologia* **41**, 11–23.

Hunter, M.L. (1990) *Wildlife, Forests and Forestry.* Prentice Hall, London.

IGM (1876–1877) *Carta delle Province Meridionali del Regno.* IGM, Firenze.

IGM (1986) *Carta Topografica di Italia.* Scala 1:50.000. Foglio n.382 'Serracapriola'. IGM, Firenze.

ISTAT (1935) *Catasto Agrario 1929*. Compartimento degli Abruzzi e Molise. Provincia di Campobasso. Fasc. 63. Istituto Poligrafico dello Stato, Roma.

ISTAT (1986) *Censimento Generale dell'Agricoltura*. Fasc.70, Campobasso. ISTAT, Roma.

Insley, H. (ed.) (1988) *Farm Woodland Planning*. Forestry Commission Bulletin 80. HMSO, London.

ISCUM (1987) *I liguri dei monti*. SAGEP, Genova.

Iturrondobeitia, J.C. and Saloña, M.I. (1990) Estudio de las comunidades de oribatidos Acari, Oribatei de varios ecosistemas de Vizcaya y una zona proxima: 2. Distribucion de abundancias y diversidad especifica. *Revue de Ecologie et Biologie du Sol* (France) **27**, 113–133.

Iturrondobeitia, J.C. and Saloña, M.I. (1991) Estudio de las comunidades de oribatidos Acari, Oribatei de varios ecosistemas de Vizcaya y una zona proxima: 4. Relacion entre las especies de oribatidos y los factores fisicoquimicos del suelo. *Revue de Écologie et Biologie du Sol* (France) **28(4)**, 443–459.

Iturrondobeitia, J.C. and Subias, L.S. (1981) Sinecologia de las comunidades de oribatidos Acarida, Oribatida del Valle de Arratia Vizcaya. *Cuadernos de Investigacion Biologica* (Bilbao) **2**, 11–25.

Jeffrey, D.J. (1987) *Soil–Plant Relationships – An Ecological Approach*. Croom Helm, London.

Jenkins, D., Watson, A. and Miller, G.R. (1967) Population fluctuations in the red grouse *Lagopus lagopus scoticus*. *Journal of Animal Ecology* **36**, 97–122.

Kelchtermans, T. (1990) Milieubeleidsplan en natuurontwikkelingsplan voor Vlaanderen. Voorstellen voor 1990–1995. (Environmental policy plan and nature development plan for Flanders. Proposals 1990–1995.) *Vlaamse Raad*. Stuk 296 (1989–1990) No.1.

Kent, M. and Wathern, P. (1980) The vegetation of a Dartmoor catchment. *Vegetatio* **43**, 163–172.

Kirby, K.J. (1988) Changes in the ground flora under plantations on ancient woodland sites. *Forestry* **61**, 317–338.

Kirby, K.J. (1990) Changes in the ground flora of a broadleaved wood within a clear fell, group fell and a coppiced block. *Forestry* **63**, 242–249.

Kirby, K.J. (1992) Accumulation of dead wood: a missing ingredient in coppicing? In: Buckley, G.P. (ed.) *Ecology and Management of Coppice Woodlands*. Chapman and Hall, London, pp.99–112.

Kirby, K.J. and May, J. (1989) The effects of enclosure, conifer planting and the subsequent removal of conifers in Dalavich Oakwood (Argyll). *Scottish Forestry* **43**, 280–288.

Kreiser, A.M., Appleby, P.G., Natkausky, J., Rippey, B. and Batterbee, R.W. (1990). Afforestation and lake acidification: a comparison of four sites in Scotland. *Philosophical Transactions of the Royal Society, London* **B 327**, 377–383.

Lacroix, G.L., Gordon, D.J. and Johnston, D.J. (1985) Effects of low environmental pH on the survival, growth and ionic composition of postemergent Atlantic Salmon (*Salmo salar*). *Canadian Journal Fisheries Aquatic Sciences* **42**, 768–775.

Lance, A.N. (1978) Territories and the food plant of individual red grouse. *Journal of Animal Ecology* **47**, 307–313.

Lance, A.N. and Lawton, J.H. (1990) *Red Grouse Population Processes. Proceedings of a*

Workshop Convened by the British Ecological Society and the Royal Society for the Protection of Birds. British Ecological Society and the Royal Society for the Protection of Birds, Sandy.

Lance, G.N. and Williams, W.T. (1967) A general theory of classificatory sorting strategies I. Hierarchical systems. *Computer Journal* **9**, 373–380.

Langan, S.J. (1989) Sea-salt induced streamwater acidification. *Hydrological Processes* **3**, 25–41.

Langslow, D.R. and Reed, T.M. (1985) Inter-year comparisons of breeding wader populations in the uplands of England and Scotland. In: Taylor, K., Fuller, R.J. and Lack, P.C. (eds) *Bird Census and Atlas Studies. Proceedings of the VIII International Conference on Bird Census and Atlas Work.* British Trust for Ornithology, Tring, pp.165–173.

Lavers, C. and Haines-Young, R.H. (in preparation) Equilibrium landscapes and their aftermath. In: Haines-Young, R.H., Cousins, S.H. and Green, D.R. (eds) *Landscape Ecology and GIS*, Taylor and Francis, London.

Lebrun, P. (1965) Contribution a l'étude écologique des Oribates de la litière dans une foret de moyenne-Belgique. *Memoires de la Société Royale des Sciences Naturelles, Belgium*, **153**, 95.

Leak, W.B. and Filip, S.M. (1977) Thirty-eight years of group selection in New England northern hardwoods. *Journal of Forestry* **75**, 641–643.

Leclercq, B. (1987) Influence de quelques pratiques sylvicoles sur la qualité des biotopes à Grand Tetras (*Tetrao urogallus*) dans le massif du Jura. *Acta Oecologia/Oecologia Generalis* **8**, 237–246.

Lee, J. (1991) Land resources, land use and projected land availability for alternative uses in the EC. In: Brouwer, F.M., Thomas, A.J. and Chadwick, M.J. (eds) *Land Use Changes in Europe.* Kluwer Academic Publishers, Dordrecht, pp.1–20.

Lindsay, R.A., Charman, D.J., Everingham, F., O'Reilly, R.M., Palmer, M.A., Rowell, T.A. and Stroud, D.A. (1988) *The Flow Country: The Peatlands of Caithness and Sutherland.* Nature Conservancy Council, Peterborough.

Lust, N. and Van Gijsel (1986) *De positie van de boomsoorten in 'Het Broek' te Blaasveld.* (The position of the tree species in 'Het Broek' at Blaasveld.) Working group SEB, University of Ghent.

MacArthur, R.H. and Wilson, E.D. (1967) *The Theory of Island Biogeography.* Princeton University Press.

Mackereth, F.J.H., Heron, J. and Talling, J.F. (1978) *Water Analysis: Some Revised Methods for Limnologists.* Freshwater Biological Association, Scientific Publication No. 36.

Mackintosh, E. (1990) Landscape considerations for various silvicultural systems in the UK. In: Gordon. P. (ed.) *Silvicultural Systems.* Institute of Chartered Foresters, Edinburgh, pp.88–109.

Macphail, R.I. (1990) Micromorphological investigation of the soils and sediments. In: Maggi, R. (ed.) *Archeologia dell'Appennino Ligure. Scavi del Castellaro di Uscio.* Istituto Internatzionale di Studi Liguri, Bordighera, pp.175–195.

Mader, D. (1984) Animal habitat isolation by roads and agricultural fields. *Biological Conservation* **29**, 81–96.

MAFF (1991) Set-aside take-up under one-year and five-year schemes. *MAFF News Release*, 17th December 456/91.

Marchettano, E. (1908) I pascoli alpini della Carnia e del Canale del Ferro. *Bull.*

Associazione Agraria Friulana (Udine), serie V, **XXV**, 387–398.

Marinelli, O. (1912) *Guida delle Prealpi Giulie.* SAF, Udine.

Marquiss, M., Newton, I. and Ratcliffe, D.A. (1978) The decline of the raven, *Corvus corax*, in relation to afforestation in southern Scotland and northern England. *Journal of Applied Ecology* **15**, 129–144.

Marquiss, M., Ratcliffe, D.A. and Roxburgh, L.R. (1985) The numbers, breeding success and diet of golden eagles in southern Scotland in relation to changes in land use. *Biological Conservation* **34**, 121–140.

Massafra, A. (1980) Orientamenti colturali, rapporti produttivi e consumi alimentari nel Molise tra '700 e '800. *Quaderni Storici* n.43. Il Mulino, Bologna, p.84.

Mather, A.S. (1991) Pressures on British forest policy: prologues to the post-industrial forest. *Area* **23**, 245–253.

McDonald, D.G. (1983) The influence of calcium on the physiological responses of the rainbow trout, *Salmo gairdneri.* I. Branchial and renal net ion and H^+ fluxes. *Journal of Experimental Biology* **102**, 123–140.

McDonald, D.G., Hobe, H. and Wood, C.M. (1980) The influence of calcium on the physiological responses of the rainbow trout, *Salmo gairdneri*, to low environmental pH. *Journal of Experimental Biology* **88**, 109–131.

Miles, J. (1981a) *Effect of Birch on Moorlands.* Institute of Terrestrial Ecology, Edinburgh.

Miles, J. (1981b) Effects of trees on soils. In: Last, F.T. and Gardiner, A.S. (eds) *Forest and Woodland Ecology.* Institute of Terrestrial Ecology Symposium, No. 8, pp.85–88.

Miles, J. (1986) What are the effects of trees on soil? In: Jenkins, D. (ed.) *Trees and Wildlife in the Scottish Uplands.* Institute of Terrestrial Ecology, Cambridge, pp. 55–62.

Miller, G.R., Jenkins, D. and Watson, A. (1966) Heather performance and red grouse populations. I. Visual estimates of heather performance. *Journal of Applied Ecology* **3**, 313–326.

Miller, H.G. (1988) Effects of forestry practices on the chemical, biological and physical properties of soils. In: Barth, H. and Hermite, P. (eds) *Scientific Basis for Soil Protection in the European Community.* Elsevier, London, pp.237–246.

Milner, N.J. and Vallaro, P.V. (1990). Effects of acidification on fish and fisheries in Wales. In: Edwards, R.W., Gee, A.S. and Stoner, J.H. (eds) *Acid Waters in Wales.* Kluwer Academic Publishers, Dordrecht.

Mitchell, P.L. and Kirby, K.J. (1989) *Ecological Effects of Forestry Practices in Long-established Woodland and their Implications for Nature Conservation.* Oxford Forestry Institute (Occasional Paper 39), Oxford.

Mirmina, E. (a cura di) (1985) Gente e territorio delle valli del Torre. Comunita' montana delle valli del Torre-Centro Friulano di Studi 'Ippolito Nievo', Udine.

Moreno, D. (1990) *Dal documento al terreno. Storia e archeologia dei sistemi agro-silvo-pastorali.* Il Mulino, Bologna.

Moreno, D., Croce, G.F. and Montanari, C. (1992) Antiche praterie apenniniche. In: Maggi R. (ed.) Archeologia preventiva lungo il percorso di un metanodotto. *Quaderni della Soprintendenza Archeologica della Liguria* 4. Chapter 6.

Moreno, D. and Croce, G.F. (1993) Storia e archeologia della risorse ambientali: il Bosco Ramasso (XIX–XX secolo). *Bollettino Ligustico*, (in press.)

Moreno, D. and Montanari, C. (1988) The use of historical photographs as a source in the study of dynamics of vegetational groups and woodlands landscape. In:

Salbitano, F. (ed.) *Human Influence on Forest Ecosystem Development in Europe.* Pitagora Editrice, Bologna, pp.371–373.

Moreno, D. and Raggio, O. (1991) The making and fall of an intensive pastoral land use system in the Eastern Liguria (XVI–XIXth C). In: Maggi, R., Nisbet, R. and Barker, G. (eds) *The Archaeology of Pastoralism in Southern Europe.* Istituto Internazionale di Studi Liguri, Bordighera, Vol.1.

Moss, D.A. (1979) Even-aged plantations as habitats for birds. In: Ford, E.D., Malcolm, D.C. and Atterson, J. *The Ecology of Even-aged Plantations.* Institute of Terrestrial Ecology, Cambridge, pp.413–427.

Musoni, F. (1915) La popolazione in Friuli. *Annali Istituto Tecnico Zanon,* Udine, serie II, anno XXXII, 1912–13, pp.4–112.

Muys, B. (1989a) Evaluation of conversion of tree species and liming as measures to decrease soil compaction in a beech forest on loamy soil. *FAO/ECE/ILO Seminar on the Impact of Mechanization of Forest Operations to the Soil.* Louvain-la-neuve, pp.341–355.

Muys, B. (1989b) *Earthworms and Litter Decomposition in the Forests of the Flemish Region.* Proceedings of the Symposium 'Invertebrates of Belgium'. KBIN, Brussels, pp.71–78.

Muys, B. and Lust, N. (1992) Inventory of the earthworm community and the state of litter decomposition in the forests of Flanders (Belgium) and its implications for forest management. *Soil Biology and Biochemistry,* (in press).

Muys, B., Lust N. and Granval, P. (1992) Effects of grassland afforestation with different tree species on earthworm communities, litter decomposition and nutrient status. *Soil Biology and Biochemistry,* (in press).

National Audit Office (1986) *Review of Forestry Commission Objectives and Achievements.* HMSO, London.

Nature Conservancy Council (1984) *Nature Conservation in Great Britain.* Nature Conservancy Council, Peterborough.

Nature Conservancy Council (1986) *Nature Conservation and Afforestation in Britain.* Nature Conservancy Council, Peterborough.

Nature Conservancy Council (1990) *Nature Conservation and Agricultural Change.* Nature Conservancy Council, Peterborough.

Nethersole-Thompson, D. and Nethersole-Thompson, M. (1979) *Greenshanks.* Poyser, Berkhamstead.

Nethersole-Thompson, D. and Watson, A. (1981) *The Cairngorms: Their Natural History and Scenery.* Melven, Perth.

Newton, I., Meek, E.R. and Little, B. (1978) Breeding ecology of the merlin in Northumberland. *British Birds* **71**, 376–398.

Newton, I., Davis, P.E. and Davis, J.E. (1982) Ravens and buzzards in relation to sheep farming and forestry in Wales. *Journal of Applied Ecology* **19**, 681–706.

Newton, I., Meek, E.R. and Little, B. (1984) Breeding season foods of the merlin *Falco columbarius* in Northumbria. *Bird Study* **31**, 49–56.

Newton, I., Meek, E.R. and Little, B. (1986) Population and breeding of Northumbrian merlins. *British Birds* **79**, 155–170.

Newton, J.P. and Rivers, M.J. (1982) Lake Vyrnwy: an example of the multiple use of rural land. *Quarterly Journal of Forestry* **76**(**2**), pp.92–102.

Noirfalise, A. (1969) La chênaie mélangée à jacinthe du domaine atlantique de l'Europe (Endymio-Carpinetum). *Vegetatio* **17**, 131–150.

Nowak, B. (1987) Untersuchungen zur Vegetation Ostliguriens (Italien). *Dissertationes Botanicae.* Band III Cramer, Berlin.

O'Connor, R.J. (1980) Population regulation in the yellowhammer. In: Oelke, H. (ed.) *Emberiza citrinella. Bird Census Work and Nature Conservation.* University of Gottingen, Gottingen, pp.190–200.

O'Connor, R.J. (1985) Behavioural regulation of bird populations: a review of habitat use in relation to migration and residency. In: Sibley, R.M. and Smith, R.H. (eds) *Behavioural Ecology: Ecological Consequences of Adaptive Behaviour.* Blackwell Scientific Publications, Oxford, pp.105–142.

O'Connor, R.J. (1986) Dynamical aspects of avian habitat use. In: Verner, J., Morrison, M.L. and Ralph, C.J. (eds) *Wildlife 2000. Modelling Habitat Relationships of Terrestrial Vertebrates.* University of Wisconsin Press, Madison, pp.235–240.

O'Connor, R.J. and Fuller, R.J. (1985) Bird population responses to habitat. In: Taylor, K., Fuller, R.J. and Lack, P.C. (eds) *Bird Census and Atlas Studies: Proceedings of the VII International Conference on Bird Census Work.* British Trust for Ornithology, Tring, pp.197–211.

O'Donald, P. (1983) *The Arctic Skua.* Cambridge University Press, Cambridge.

O'Gorman, (1845) *The Practice of Angling, Particularly as Regards Ireland.* William Curry and Co., Dublin (2 vols.)

Orford, N. (1973) Breeding distribution of the twite in central Britain. *Bird Study* **20**, 51–62 and 121–126.

Ormerod, S.J., Wade, K.R. and Gee, A.S. (1987). Macro-floral assemblages in upland Welsh streams in relation to acidity and their importance to invertebrates. *Freshwater Biology* **18**, 545–558.

Ostermann V. (1940) *Le arti e le tradizioni popolari in Italia.* Istituto Edizioni Accademiche, Udine.

Ovington, J.D. (1953) Studies of the development of woodland conditions under different tree species. *Journal of Ecology* **41**, 13–34.

Ovington, J.D. (1955) Studies of the development of woodland conditions under different trees. III. The ground flora. *Journal of Ecology* **43**, 1–21.

Ovington, J.D. (1956) The composition of tree leaves. *Forestry* **19**, 22–28.

Page, G. (1968) Some effects of conifer crops on soil properties. *Commonwealth Forestry Review* **47**, 52–62.

Parr, R. (1990) *Moorland Birds and Their Predators in Relation to Afforestation.* NCC Chief Scientist Directorate commissioned research report No.1081. [ITE Final Report on contract HF3/08/24 to NCC.]

Peachey, (1980) The Conservation of Butterflies in Bernwood Forest. Unpublished, Nature Conservancy Council, Newbury.

Pepe, R. (1811) Dello stato e conservazione de'boschi della Provincia di Molise. Atti dell'Istituto Reale di *Incoraggiamento alle Scienze Naturali.* Napoli, pp.211–214.

Pepe, R. (1834) Qualche parola sulle cose rustiche della Provincia di Molise. I Boschi. *Annali Civili.* Fascicolo VII. Gen-Feb. Tipografia del Real Ministero di Stato degli Affari Interni nel Reale Albergo de' Poveri, Napoli, p.21.

Pepe, R. (1844) Poche osservazioni di economia campestre intorno a'pascoli boscosi della Provincia di Molise. *Annali Civili.* Fascicolo LXIX. Tipografia del Real Ministero di Stato degli Affari Interni nel Reale Albergo de' Poveri, Napoli, pp.8–17.

Petch, A.M. (1965) Leaf trees as soil improvers. *Quarterly Journal of Forestry* **59**(**4**), 318–322.

Peterken, G.F. (1974) A method for assessing woodland flora for conservation using indicator species. *Biological Conservation* **6**, 239–245.

Peterken, G.F. (1976) Long-term changes in the woodlands of Rockingham Forest and other areas. *Journal of Ecology* **64**, 123–146.

Peterken, G.F. (1977a) Habitat conservation priorities in British and European woodlands. *Biological Conservation* **11**, 223–236.

Peterken, G.F. (1977b) General management principles for nature conservation in British woodland. *Forestry* **50**, 27–48.

Peterken, G.F. (1981) *Woodland Conservation and Management.* Chapman and Hall, London.

Peterken, G.F. (1987) Natural features in the management of upland conifer forests. *Proceedings of the Royal Society of Edinburgh* **93B**, 223–234.

Peterken, G.F. (1992) Coppices in the lowland landscape. In: G.P. Buckley (ed.) *Ecology and Management of Coppice Woodlands.* Chapman and Hall, London, pp.3–17.

Peterken, G.F. and Game, M. (1981) Historical factors affecting the distribution o *Mercurialis perennis* in central Lincolnshire. *Journal of Ecology* **69**, 781–796.

Peterken, G.F. and Game, M. (1984) Historical factors affecting the number and distri bution of vascular plant species in the woodlands of central Lincolnshire. *Journal o Ecology* **72**, 155–182.

Peterken, G.F. and Hughes, F.M.R. (1990) The changing lowlands. In: Bayliss-Smith, T and Owens, S.E. (eds) *Britain's Changing Environment from the Air.* Cambridg University Press, Cambridge, pp.48–76.

Pettenella, D. (1988) Gestione delle risorse forestali e tutela dell'ambiente in Molis *Molise Economico*, 1/2/3. Campobasso.

Piccioli, L. and Piccioli, E. (1923) Alpicoltura. In: *Nuova Enciclopedia Agraria Italian* Torino, pp.11–20.

Picozzi, N. and Wier, D. (1974) Breeding biology of the buzzard in Speyside. *British Bir* **67**, 199–210.

Picozzi, N. and Wier, D. (1976) Dispersal and causes of death of buzzards. *British Bir* **69**, 193–201.

Pienkowski, M.W., Stroud, D.A. and Reed, T.M. (1987) Problems in maintaini breeding habitat, with particular reference to peatland waders. *Wader Study Gro Bulletin* **49**, 95–101.

Pigott, C.D. (1990) The influence of evergreen coniferous nurse-crops on the field layer two woodland communities. *Journal of Applied Ecology* **27**, 448–459.

Pollard, E. (1973) Hedges. VII. Woodland relic hedges in Huntingdon and Peterborou *Journal of Ecology* **61**, 343–352.

Pollard, E. (1982) Monitoring butterfly abundance in relation to the management o nature reserve. *Biological Conservation* **24**, 317–328.

Potter, C., Burnham, P., Edwards, A., Gasson, R. and Green, B. (1991) *The Diversion Land: Conservation in a Period of Farming Contraction.* Routledge, London.

Prost, B. (1977) *Il Friuli. Regione di incontri e scontri.* Geneve.

Pyatt, D.G. (1990) Long-term prospects for forests on peatland. *Scottish Forestry* **44**, 19–

Pyatt, D.G. and Craven, M.M. (1979) Soil changes under even-aged plantations. In: Fo E.D., Malcolm, D.C. and Atterson, J. (eds) *The Ecology of Even-aged Plantatio* Institute of Terrestrial Ecology, Cambridge, pp.369–388.

Rackham, O. (1980) *Ancient Woodland. Its History, Vegetation and Uses in Engla* Edward Arnold, London.

Rackham, O. (1986) *The History of the Countryside.* Dent, London.

Ratcliffe, D.A. (1968) An ecological account of Atlantic bryophytes in the British Isles. *New Phytologist* **67**, 365–439.

Ratcliffe D.A. (1976) Observations on the breeding of the golden plover in Great Britain. *Bird Study* **23**, 63–116.

Ratcliffe, D.A. (ed.) (1977) *A Nature Conservation Review*, Vol.1. Cambridge University Press, Cambridge.

Ratcliffe D.A. (1986) The effects of afforestation on the wildlife of open habitats. In: Jenkins, D. (ed.) *Trees and Wildlife in the Scottish Uplands.* Institute of Terrestrial Ecology, Banchory, pp.46–54.

Ratcliffe, D.A. and Thompson, D.B.A. (1988) The British uplands: their ecological character and international significance. In: Usher, M.B. and Thompson, D.B.A. (eds) *Ecological Change in the Uplands.* Blackwell Scientific Publications, Oxford, pp.9–36.

Rehfüss, K.E., Rosch, I. and Makeschin, F. (1990) Short rotation plantations in Central Europe: nutrition and influences on soil factors. In: *Proceedings of the XIXth IUFRO World Congress.* Montréal. Div. 1, Vol.1, 214–220.

Renewable Energy Enquiries Bureau (REEB) (1991) *Arable Coppice.* REEB, Harwell.

Rennie, P.J. (1962) Some long-term effects of tree growth on soil productivity. *Commonwealth Forestry Review* **41**, 209–213.

Ripple, W.J., Johnson, D.H., Hershey, K.T. and Meslow, E.C. (1991) Old growth and mature forests near spotted owl nests in western Oregon. *Journal of Wildlife Management* **55**, 316–318.

Roberts, A.S., Russell, C., Walker, G.J. and Kirby, K.J. (1992) Regional variation in the origin, extent and composition of Scottish woodland. *Botanical Journal of Scotland*, (in press).

Royal Society for the Protection of Birds (1985) *Forestry in the Flow Country – The Threat to Birds.* Royal Society for the Protection of Birds, Sandy.

Salbitano F. (1987) Vegetazione forestale ed insediamento del bosco in campi abbandonati in un settore delle Prealpi Giulie (Taipana-Udine). *Gortania. Atti del Museo Friulano di Storia Naturale* **9**, 183–144.

Salbitano, F. (ed.) (1988) *Human Influence on Forest Ecosystems Development in Europe.* Pitagora Editrice Bologna, Bologna.

Saloña, M.I. and Iturrondobeitia, J.C. (1988a) Estudio de las comunidades de oribatidos Acari, Oribatei de varios ecosistemas de Vizcaya y una zona proxima: 1. Sistematica y listado faunistico. *Kobie* (Bilbao) **17**, 79–92.

Saloña, M.I. and Iturrondobeitia, J.C. (1988b) Estudio de las comunidades de oribatidos Acari, Oribatei de varios ecosistemas de Vizcaya y una zona proxima: 3. Analisis comparado de las afinidades cenoticas e interespecificas. *Revue de Ecologie et Biologie du Sol* **27**, 185–203.

Selan, U. (1906) L'industria zootecnica nella Slavia italiana. *Bull. Associazione Agraria Friulana*, serie V, vol.XXIII, pp.111–124.

Selman, P. (1990) Forestry and land-use planning: a case for indicative strategies. *Arboricultural Journal* **14**, 53–59.

Selman, P. (1993) Landscape ecology and countryside planning. *Journal of Rural Studies*, (in press).

Selman, P. and Doar, N.R. (1992) An investigation of the potential for landscape ecology to act as a basis for rural land use plans. *Journal of Environmental Management* **35**, 281–299.

Senner, S.E. and Marshall, A.H. (1984) Conservation of nearctic shorebirds. In: Burger, J. and Olla, B.L. (eds) *Behaviour of Marine Animals*, Vol.5, pp.379–421.

Sharrock, J.T.R. (1976) *The Atlas of Breeding Birds in Britain and Ireland.* Poyser, Calton.

Sherfield, Lord (1980) *Scientific Aspects of Forestry.* 2nd report of the House of Lords Select Committee on Science and Technology, HMSO, London.

Siegel, S. (1956) *Nonparametric Statistics for the Behavioural Sciences.* McGraw-Hill, Kogakusha Company, Tokyo.

Siepel, H., Meijer, J., Mabelis, A.A. and Den Boer, M.H. (1989) A tool to assess the influence of management practices on grassland surface macrofaunas. *Journal of Applied Entomology* **108**, 271–290.

Simmons, E.A. and Buckley, G.P. (1992) Ground vegetation under planted mixtures of trees. In: Cannell, M.G.R., Malcolm, D.C. and Robertson, P.A. (eds) *The Ecology of Mixed-species Stands of Trees.* Blackwell Scientific Publishers, Oxford, pp.211–232.

Skartveit A. (1981) Relationship between precipitation chemistry, hydrology and runoff acidity. *Nordic Hydrology* **12**, 76–80.

Smith, K.W., Averis, B. and Martin, J. (1987) The breeding bird community of oak plantations in the Forest of Dean, southern England. *Acta Oecologia/Oecologia Generalis* **8**, 209–217.

Sneath, P.H.A. and Sokal, R.S. (1973) *Numerical Taxonomy.* W.H. Freeman and Co., San Francisco.

Soikkeli, M. (1967) Breeding cycle and population dynamics in the dunlin (*Calidris alpina*). *Annalles Zoologici Fennici* **4**, 158–198.

Soikkeli, M. (1970) Dispersal of dunlin (*Calidris alpina*) in relation to sites of birth and breeding. *Ornis Fennica* **47**, 1–9.

Sokal, R.R. and Rohlf, F.J. (1979) *Biometria.* Ed. Blume.

Spellerberg, I.F. (1988) Ecology and management of reptile populations in forests. *Quarterly Journal of Forestry* **82**, 99–109.

Spencer, J.W. and Kirby, K.J. (1992) An inventory of ancient woodland for England and Wales *Biological Conservation* **62**, 77–93.

Sprengel, U. (1971) Die Wanderherdenwirtschaft im mittel und südost italienischen Raum. *Marburger Geographische Schriften.* Heft 51. Nolte, Marburg-Lahn, p.54.

Steele R.C. and Peterken G.F. (1982) Management objectives for broadleaved woodland conservation. In: Malcolm, D.C., Evans, J. and Edwards, P.N. (eds) *Broadleaves in Britain.* Institute of Chartered Foresters, Edinburgh, pp.91–103.

Stevens, P.A. (1987) Throughfall chemistry beneath Sitka of four ages in Beddgelert Forest, North Wales, UK. *Plant and Soil* **101**, 291–294.

Stoffel, J.W. (1985) Ruimtelijke variatie in een jong populierenbos. (Spatial variation in a young poplar plantation.) *Populier* **22**(**4**), 76–79.

Stoner, J.H. and Gee, A.S. (1985) Effects of forestry on water quality and fish in Welsh rivers and lakes. *Journal of the Institute of Water Engineers and Scientists* **39**, 27–45.

Stroud, D.A. and Reed, T.M. (1986) The effects of plantation proximity on moorland breeding waders. *Wader Study Group Bulletin*, **46**, 25–28.

Stroud, D.A., Reed, T.M., Pienkowski, M.W. and Lindsay, R.A. (1987) *Birds, Bogs and Forestry.* Nature Conservancy Council, Peterborough.

Stroud, D.A., Reed, T. and Harding, N.J. (1990) Do moorland birds avoid plantation edges? *Bird Study* **37**, 177–186.

Svardson, G. (1949) Competition and habitat selection in birds. *Oikos* **1**, 157–174.

Taffetani, F. (1991) Il litorale nord dell'antica Capitanata. *Almanacco del Molise*, Vol.I. Edizioni Enne, Campobasso.

Thirsk, J. (1957) *English Peasant Farming*. Routledge and Kegan Paul, London.

Thom, V.M. (1986) *Birds in Scotland.* Poyser, Calton.

Thomas, R.C. (1987) "The Historical Ecology of Bernwood Forest." PhD thesis, Oxford Polytechnic, Oxford.

Thomas, R.C. (1988) Historical ecology of Bernwood Forest Nature Reserve 1900–1981. In: Kirby, K.J. and Wright, F.J. (eds) *Woodland Conservation and Research in the Clay Vale of Oxfordshire and Buckinghamshire.* Nature Conservancy Council, Peterborough, pp.20–29.

Thompson, D.B.A., Thompson, P.S. and Nethersole-Thompson, D. (1986) Timing of breeding and breeding performance in a population of greenshanks, *Tringa nebularia. Journal of Animal Ecology* **55**, 181–199.

Thompson, D.B.A., Stroud, D.A. and Pienkowski, M.W. (1988) Afforestation and upland birds: consequences for population ecology. In: Usher, M.B. and Thompson, D.B.A. (eds) *Ecological Change in the Uplands.* Blackwell Scientific Publications, Oxford, pp.237–259.

Tompkins, S. (1989) *Forestry in Crisis: The Battle for the Hills.* Christopher Helm, London.

Turnpenny, A.W.H., Sadler, K., Ashton, R.J., Milner, A.G.P. and Lynam, S. (1987) The fish populations of some streams in Wales and Northern England in relation to acidity and associated factors. *Journal of Fish Biology* **31**, 415–434.

Ullmann, R. (1967) *Der Nordwestliche Apennin. Kulturgeographische Wandlungen seit des 18 Jahhunderts.* Freiburger Geographiche Arbeiten, 2, H. F. Schulz, Freiburg.

Ulrich, B. (1981) Destabilisierung von Waldökosystemen durch Biomassenutzung. *Forstarchiv* **52(6)**, 199–203.

Ulrich, B. (1983) Stabilität von Waldökosystemen unter dem Einfluss des 'sauren Regens'. *Allgemeine Forstzeitschrift*, **26/27**, 670–677.

Usher, M.B. and Gardner, S.M. (1988) Animal communities in the uplands: how is naturalness influenced by management? In: Usher, M.B. and Thompson, D.B.A. (eds) *Ecological Change in the Uplands.* Blackwell Scientific Publications, Oxford, pp.75–92.

Usher, M.B. and Thompson, D.B.A. (eds) (1988) *Ecological Change in the Uplands.* Blackwell Scientific Publications, Oxford.

Valussi, G. (1971) La popolazione della regione. Il popolamento. *Enciclopedia monografica del Friuli-Venezia Giulia* **1**, 759–768.

Van Miegroet, M. (1990) The nature of forest degradation. In: *'Rehabilitation of Degraded Forest Ecosystems' Proceedings of the International Meeting of Professors of Silviculture.* IUFRO, Chania.

Van Hees, A.F.M. (1978) *Bosbeheer, vegetatie en avifauna in enkele bosgebieden in Midden-Brabant.* (Forest management, vegetation and avifauna in some forests in Midden-Brabant.) De Dorschkamp, Wageningen. Report No.159.

Vannier, G. (1979) Relations trophiques entre la microfaune et la miceoflore du sol; aspects qualitatifs et quantitatifs. *Bulletin of Zoology* **46**, 343–361.

Venturini, S. and, Tunis G. (1988) Nuovi dati ed interpretazioni sulla tettonica del settore meridionale delle Prealpi Giulie e della regione al confine tra l'Italia e la Jugoslavia. *Gortania. Atti Museo Friulano di Storia Naturale* **10**, 5–34.

Verlinden, A. (1987) *Ekologische achtergronden van het huidig bosbeheer in Vlaanderen.*

(Ecological backgrounds of the current forest management in Flanders.) Working group SEB, University of Ghent. Report No.12.

Virkkala, R. (1991) Population trends of forest birds in a Finnish Lapland landscape of large habitat blocks: consequences of stochastic environmental variation or regional habitat alteration? *Biological Conservation* **56**, 223–240.

Wallace, H.L., Good, J.E.G. and Williams, T.G. (1992) The effects of afforestation on upland plant communities: an application of the British National Vegetation Classification. *Journal of Applied Ecology* **29**, 180–194.

Ward, N., Marsden, T. and Munton, R. (1990) Farm landscape change: trends in upland and lowland England. *Land Use Policy*, October, 291–302.

Warren, M.S. (1987) The ecology and conservation of the heath fritillary butterfly, *Mellicta athalia*. III. Population dynamics and the effects of habitat management. *Journal of Applied Ecology* **24**, 499–513.

Warren, M.S. and Fuller, R.J. (1990) *Woodland Rides and Glades*. Nature Conservancy Council, Peterborough.

Warren, S.C. (1989) *Acidity in United Kingdom Fresh Waters*. Second report of the UK Acid Waters Review Group, HMSO, London.

Watkins, C. (1984) The use of grant aid to encourage woodland planting in Great Britain. *Quarterly Journal of Forestry* **78**, 213–224.

Watkins, C. (1986) Recent changes in government policy towards broadleaved woodland. *Area* **18**, 117–122.

Watkins, C. (1990) *Woodland Management and Conservation*. David and Charles, Newton Abbot.

Watkins, C. (1991) *Nature Conservation and the New Lowland Forests*. Nature Conservancy Council, Peterborough.

Watson, A. and Moss, R. (1980) Advances in our understanding of the population dynamics of red grouse from a recent fluctuation in numbers. *Ardea* **68**, 103–111, 113–119.

Watson, A., Moss, R. and Parr, R. (1984) Effects of food enrichment on numbers and spacing behaviour of red grouse. *Journal of Animal Ecology* **53**, 663–678.

Watson, J., Langslow, D.R. and Rae, S.R. (1987) *The Impact of Land-use Changes on Golden Eagles* Aquila chrysaetos *in the Scottish Highlands*. Nature Conservancy Council, Peterborough.

Watt, A.S. (1925) On the ecology of British beechwoods with specific reference to their regeneration. *Journal of Ecology* **13**, 27–73.

Wauthy, G. (1982a) Revue critique des relations entre la faune, la matière organique et les microorganismes dans les horizons organiques forestiers. *Agronomie* **2**, 667–675.

Wauthy, G. (1982b) Synecology of forest soil oribatid mites of Belgium Acari, Oribatida. III. Ecological groups. *Acta Oecologica, Oecologica Generalis* **3**, 469–494.

Wauthy, G., Noti, M.Y. and Dufrene, M. (1989) Geographic ecology of soil oribatid mites in deciduous forests. *Pedobiologia* **33**, 399–416.

Wheeler, P.T. (1984) A survey of woodland change in Nottinghamshire, 1920–1980. *East Midland Geographer* **8**, 134–147.

Wheeler, P.T. (1988) Twentieth century change in woodland location, area and type: a Nottinghamshire example. In: Salbitano, F. (ed.) *Human Influence on Forest Ecosystems Development in Europe*. Pitagora Editrice Bologna, Bologna, pp.179–188.

Whitney, G.G. and Foster, D.R. (1988) Overstorey composition and age as determinants of the understorey flora of woods of central New England. *Journal of Ecology* **76**, 867–876.

Wiens, J.A. (1984) On understanding a non-equilibrium world: myth and reality in community patterns and processes. In: Strong, D.R., Simberloff, D., Abele, L.G., and Thistle, A.B. (eds) *Ecological Communities: Conceptual Issues and the Evidence.* pp.439–457.

Wiens, J.A. (1989a) *The Ecology of Bird Communities, Vol.1: Foundations and Patterns.* Cambridge University Press, Cambridge.

Wiens, J.A. (1989b) *The Ecology of Bird Communities, Vol.2: Processes and Variations.* Cambridge University Press, Cambridge.

Wiklander, L. (1975). The role of neutral salts in the ion exchange between acid precipitation and soil. *Geoderma* **14**, 93–105.

Winter, M., Richardson, C., Short, C. and Watkins, C. (1990) *Agricultural Land Tenure in England and Wales.* Royal Institution of Chartered Surveyors, London.

Wittich, W. (1948) *Die heutigen Grundlagen der Holzartenwahl.* Verlag Schaper, Hannover.

Wittich, W. (1963) Bedeutung einer leistungfähigen Regenwurmfauna unter Nadelwald für Streuzersetzung, Humusbildung und allgemeine Bodendynamik. *Schriftenreihe der forstliche Fak. Univ. Göttingen* **30**, 60.

Wood, C. (1989) The physiological problems of fish in acid waters. In: Morris, R., Taylor, E., Brown, D., and Brown, J. (eds) *Acid Toxicity and Aquatic Animals.* Cambridge University Press, Cambridge, pp.125–152.

Woodruffe-Peacock, E.A. (1918) A fox-covert study. *Journal of Ecology* **6**, 110–123.

Wright, R.F., Norton, S.A., Brakke, D.F. and Frogner, T. (1988) Experimental verification of episodic acidification of freshwaters by sea salts. *Nature* **334**, 422–424.

Yeates, G.W. (1988) Earthworm and Enchytraeid populations in a 13-year old agroforestry system. *New Zealand Journal of Forestry Science* **18**(3), 304–310.

Index